Cambridge

September 1977.

RESOURCE DEVELOPMENT IN SEMI-ARID LANDS

RESOURCE DEVELOPMENT IN SEMI-ARID LANDS

A ROYAL SOCIETY DISCUSSION
ORGANIZED BY
SIR JOSEPH HUTCHINSON, F.R.S., A. H. BUNTING,
A. R. JOLLY AND H. C. PEREIRA, F.R.S.

LONDON
THE ROYAL SOCIETY
1977

Printed in Great Britain for the Royal Society
at the
University Press, Cambridge

ISBN 0 85403 088 3

First published in *Philosophical Transactions of the Royal Society of London*, series B, volume 278 (no. 962,) pages 437–614

Published by the Royal Society
6 Carlton House Terrace, London SW1Y 5AG

Preface

The widespread droughts of the early 1970s directed world attention to the problems of human ecology in the vast areas with marginal rainfall that are to be found in all the tropical and sub-tropical continents of the world. These problems were not created by drought. Drought is but the dry extreme of the natural climate variability of semi-arid areas. The problems arise from the pressure of human and domestic animal populations on limited resources, and they have been aggravated by the rapid expansion of populations in recent years.

Experience of exploiting the resources of semi-arid lands has been gained in many parts of the tropical and sub-tropical world. There are areas where exploitation has led to serious degradation of the resource base, followed by imaginative and successful rehabilitation, as well as areas where the degradation continues, and there is at the present time no prospect of reversing the process.

With a view to promoting an exchange of information on current changes in the ecosystems of dry countries, on the responsibility of man for degradation of the ecosystems and for their rehabilitation, and on the prospects of mobilizing such further resources as exist, the Royal Society sponsored a two-day Discussion Meeting on 'Resource development in semi-arid lands'. The meeting was held on 17 and 18 March 1976. The papers presented and the salient points of the discussions that ensued are published in this volume.

The scope of the discussion was international. The most pressing problems at the time were in the drought stricken areas on the margins of the Sahara, but the arid areas of India have also experienced crippling drought in recent years. The climatic circumstances in the Sahara and in India resemble those of Australia, southern Africa, and the United States, and any survey of the problems and of possible solutions must take account of the scientific work that has been carried out in those areas.

It is in the nature of these marginal areas that agriculturally disastrous seasons are within the normal range of climatic variation. Such seasons must be accepted as something to be foreseen and planned for, and not as an 'act of God' to be met by international charity. The difficulty of planning for drought conditions and hence of preventing famine has been increased by the great successes of science and technology in the last quarter century which have been in the control of parasitic and nutritional disease. Improvement in health has led to great increases in human and animal populations, and these have been maintained by fuller exploitation of the environment. Thus the pressure at the margin has increased, and what was a shortage that could be met by encroachment on unused resources becomes a famine because no unused resources remain.

These problems are not new, though they have been aggravated in recent years, and the meeting was planned to facilitate the exchange of information and experience gathered over the years in the major semi-arid regions of the world. The result was an impressive evaluation of the state of the semi-arid regions of the world, and of the prospects for their inhabitants. The papers are published as a contribution to knowledge on which the wise use of resources must be based.

J. B. Hutchinson

CONTENTS

[Two plates]

CONTENTS

Phil. Trans. R. Soc. Lond. B. **278**, 439–455 (1977) [439]
Printed in Great Britain

Rainfall: scarce resource in 'opportunity country'

By R. C. Rainey, F.R.S.
Centre for Overseas Pest Research, Ministry of Overseas Development, London W8 5SJ

The rainfall of semi-arid lands represents a major resource in whose more effective utilization both contemporary and traditional technologies have increasing parts to play. Though infrequent, rains can be both heavy and widespread even in regions normally arid, as a result of major disturbances of global systems of winds and weather which are typically recognizable in real time in the course of routine synoptic weather analysis.

Introduction

Rainfall is a key resource which even in arid lands can rarely be regarded as totally negligible in amount, and can probably never be regarded as random in incidence. The spatial and temporal distribution of rainfall have implications for development in arid and semi-arid lands which are considered particularly appropriate for emphasis and discussion at this meeting. Attention is also drawn to the authoritative review of the very important topic of climatic change which is being presented elsewhere today (Mason 1976). The central point of the present paper is illustrated by table 1, which summarizes the records of the four driest fully documented rainfall stations in the World Meteorological Organization's current set of world-wide climatological normals – all four in the Sahara – together with corresponding figures from arid areas of Arabia and Indo-Pakistan. It will be noted that even Wadi Halfa, with its mean annual rainfall of 3 mm from a 24-year record, has still had a July with 29 mm of rain and an October with 14 mm; and the other five stations have all recorded individual totals of 25–330 mm of rain in a single month.

Table 1. Heavy rains in arid lands (millimetres)

		mean annual rainfall	wettest month on record	next wettest month
Sudan	Wadi Halfa 1937–60	3	29 (July)	14 (October)
	Abu Hamed 1931–60	17	110 (July)	48 (August)
Algeria	Aoulef 1931–60	11	25 (January)	20 (Aug. & Nov.)
Chad	Faya-Largeau 1941–60	16	48 (June)	31 (August)
South Yemen	Aden 1941–60	39	55 (November)	47 (September)
Pakistan	Jacobabad 1931–60	99	330 (July)	129 (August)

W.M.O. (1967) Climatological normals for 1931–60.

30-2

Rain in the southeastern Sahara: some general considerations

Evidence on the spatial extent of individual falls of rain is less readily available, but areas of the order of hundreds of thousands of square kilometres can be involved, indicating the scale of a resource usually exploited only by nomads and by migrant insects (the latter with a behaviour pattern which can enable them to reach the rain areas by using the wind-systems involved). Thus 28 of Wadi Halfa's peak month's total of 29 mm, in July 1950, fell during three consecutive days (19 on one of them); rain not only fell within this same period at both neighbouring stations (Station 6, 180 km to the SE, and Dongola, 320 km to the SSW) but must also have occurred at about this time over extensive uninhabited areas around Wadi Howar, some 700 km away to the SW, where widespread desert locust breeding (for which soil moisture equivalent to about 20 mm of rain is essential (Magor 1962)) was subsequently found to have taken place. Furthermore, at about this same time a French airline pilot, with considerable experience on trans-Saharan routes, commented, during debriefing in Nairobi, on having encountered rain in this same general area, further north into the Sahara than he had ever previously seen.

Making better use of such rains begins with recognizing, in real time, the situations which produce them. The low average rainfall of the great arid zones represents an inescapable effect of the global circulation of the atmosphere; this circulation is ultimately driven by the differential solar heating of tropical and polar latitudes; and these exceptional rains result from major disturbances of the normal circulation. Thus the major arid zones are normally dominated by air which has been dried out by a process of ascent in tropical latitudes, to about the level of the base of the stratosphere at around 15 km, during which its temperature will have fallen by some 100 °C with a consequent loss of nearly all its substantial initial water-vapour content (falling out as tropical rains) and a gain of the corresponding latent heat. Much of this air descends (and is further heated by compression) in the sub-tropical anticyclonic belts, with semi-permanent cells (characterized by low rainfall) like the familiar Azores anticyclone and the corresponding St Helena anticyclone in the South Atlantic in oceanic areas as well as others in continental regions. This upper-wind circulation was originally inferred by George Hadley (1735), an early Fellow of this Society, and reported in our *Philosophical Transactions* some 240 years ago, but it is only quite recently that convincing direct evidence of this circulation has been provided, by the extension of regular upper-air observations undertaken from the 1950s to meet the requirements of high-flying jet aircraft.

The aridity of these lands may therefore be somewhat crudely envisaged as being continuously replenished from above – in such a manner as to dispose unambiguously of the possibilities which have been suggested, for example in the Schwartz scheme of half a century ago for the northern Kalahari, for significantly increasing rainfall by increasing evaporating surfaces, such as by the construction of reservoirs or by afforestation. A recent demonstration of the insignificance of this effect has been provided by the Aswan high dam, where the evaporation from Lake Nasser has resulted simply in an increased local incidence of early-morning mist in the winter months – by overnight humidification of the lowest tens of metres of the atmosphere, effectively only skin-deep, and rapidly dispersed by convective mixing after sunrise. This point of course in no way weakens the case for afforestation on other grounds such as for combatting soil-erosion.

On the weather maps of the continental tropics the very low dew-points characteristic of air with a recent history of such desiccation and descent contrast conspicuously with the high dew-points of air recently off the tropical oceans and rain forests. A particularly impressive demonstration of this contrast is provided by some of the continental stretches of the intertropical front (i.t.f.); this is another feature of the global circulation, at which surface winds from the southern hemisphere can be envisaged as meeting winds from the northern hemisphere, and incidentally found to represent a dominating feature in the atmospheric environment of tropical migrant insects.

Figure 1 shows a set of observations made with a Pilatus Porter agricultural aircraft equipped with a Decca Doppler radar system for precision wind-finding; these were made during one of a series of research projects of the Ciba-Geigy Agricultural Aviation Research Unit at Cranfield, to whose Director, Vernon Joyce, we have been greatly indebted for opportunities of participation. Here can be seen northeasterlies which have previously been dried out as just described, still with dew-points of only 4–5 °C, meeting southerlies with relatively recent oceanic tracks which have given dew-points of 17–19 °C, at a frontal discontinuity with the main zone of transition as located by the wind-shift no more than a few kilometres thick. The effect of this humidity contrast could be seen in the sweat which appeared on the back of one's hand within a minute or two of flying southwards through the front – whose very existence, incidentally, is still said to be doubted by some tropical meteorologists in other regions.

These same observations conveniently introduce and illustrate the very important process of wind-convergence, most simply envisaged as giving a net inflow of wind across a closed perimeter, and more easily visualized from figure 2, which shows the size of the mean inflow or outflow component at right angles to each side of the rectangular flight-pattern. Convergence at low levels (where most of the atmosphere's water vapour is to be found) necessarily means ascent of air, and is a necessary though not sufficient condition for the production of rain.

The detailed three-dimensional structure of the intertropical front in tropical Africa is however not yet completely understood, with the rains commonly well to the south of the surface front – within a hundred kilometres on this particular morning, but often 200–500 km away (Osman & Hastenrath 1969). The exceptional rains of July 1950 at Wadi Halfa were associated with the i.t.f. being unusually far north, reaching 25° N at 12 h G.M.T. on 24 July, to the north of Aswan, some 900 km to the north of its average position at this time of year and 350 km north of Wadi Halfa. (A similarly exceptional northward surge of the i.t.f. appears to have been associated with exceptional desert rains at Kufra oasis at 23° N in Libya in August 1952 – Pedgley 1974.) I am indebted to the Sudan Meteorological Department for information and guidance on these points, as on many others in the course of the past quarter of a century.

The seasonal movements of the intertropical convergence zone (i.t.c.z., within which the i.t.f. is one particular manifestation), following the Sun, are associated in general terms with the seasonal rains of the tropics, and in turn with regular seasonal migrations of pastoral human populations such as the Rizeiqat of the western Sudan. The Rizeiqat are a Baggara tribe estimated to include some 180 000 nomads, with about $1\frac{1}{2}$ million cattle, migrating regularly between winter grazing areas to the south of the Bahr el Arab and summer grazing areas some hundreds of kilometres away to the north, in central and northern Darfur. A recent consultant's report (for which I am greatly indebted to Hunting Technical Services) on an F.A.O./U.N.D.P. savanna development project, referring to these grazing practices of the Rizeiqat, described their 'nomadic system (as) a great asset to the nation...the nomadic producers of southern

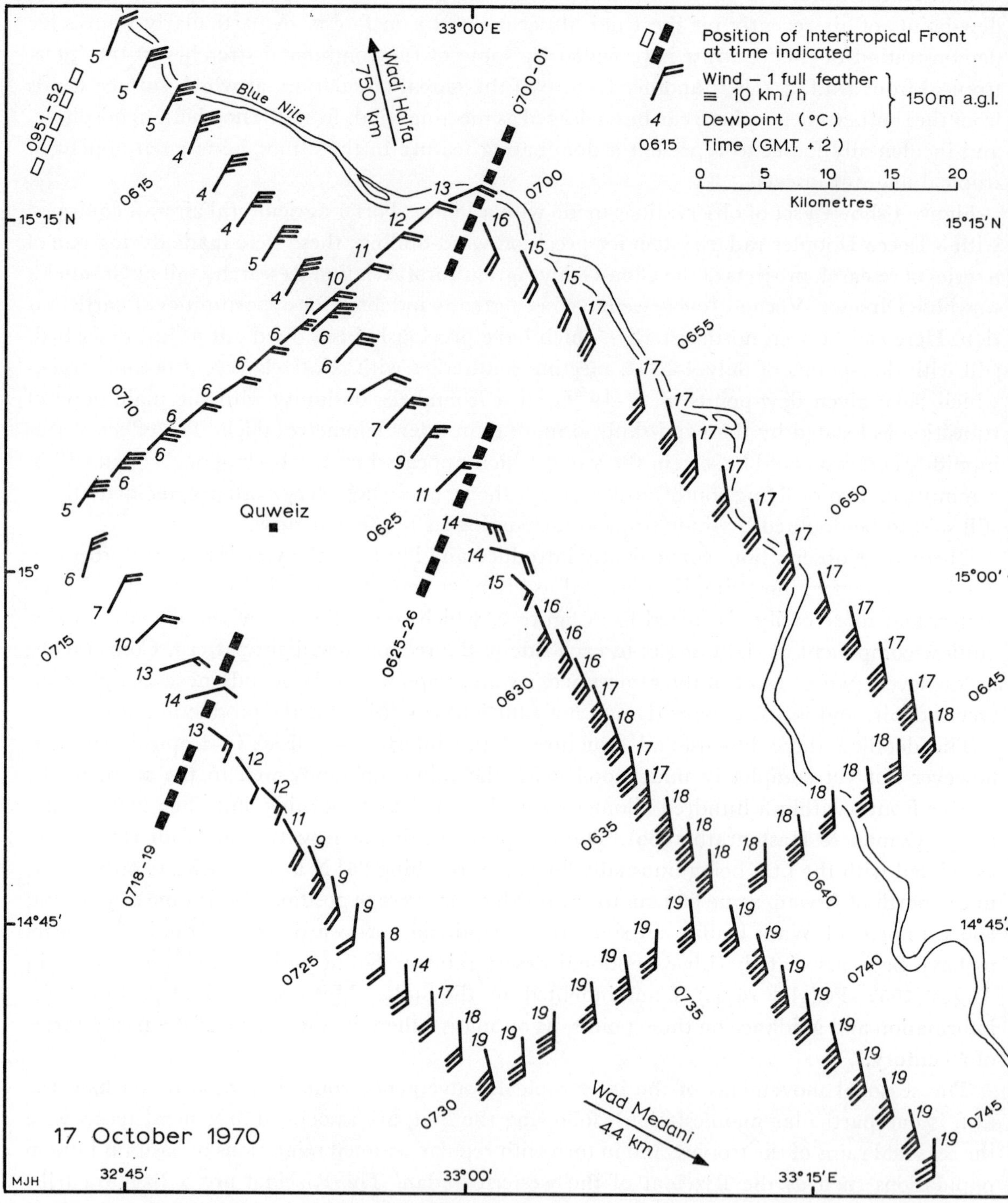

FIGURE 1. Air-mass contrast and wind-convergence at the intertropical front. Dry northeasterly winds with a history of subsidence in the subtropical anti-cyclone belt meet humid southerly monsoon air at the sharply defined intertropical front with marked convergence, associated with rain at the time around Wad Medani.

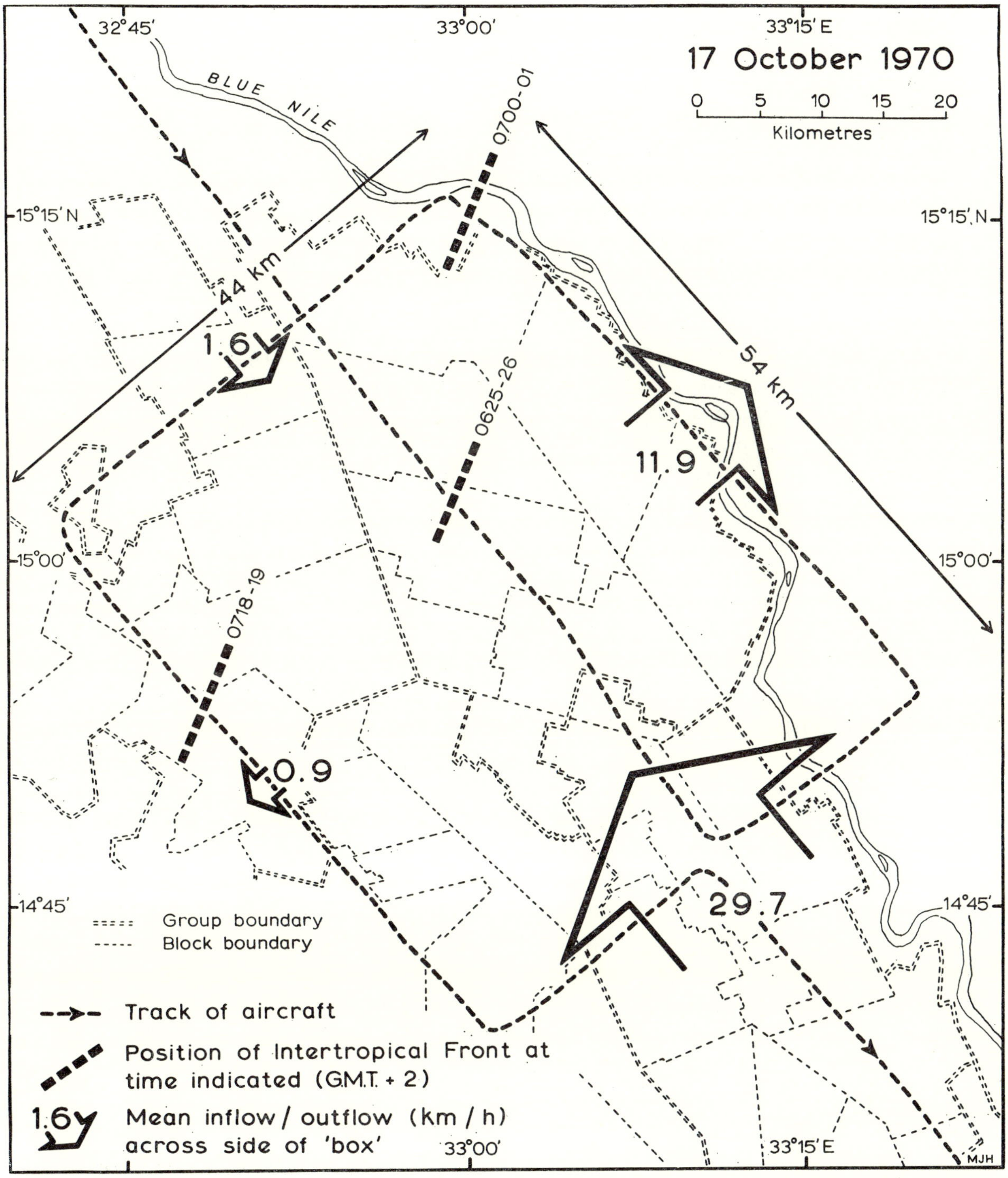

FIGURE 2. Estimation of wind-convergence. Low-level wind-convergence inevitably involves ascent of air and is essential for the production of rain; it can also concentrate flying insects. If around a closed horizontal pattern of area A the wind experienced over one small element, of length δl has an inward component of velocity V in a direction at right angles to the element, the average convergence over the area of the pattern is $\Sigma V\delta l/A$. For the observations of figure 1, treated as effectively instantaneous, this expression becomes $[44(29.7+1.6)-54(11.9+0.9)]/44\times 54$, giving convergence at an estimated rate of 0.29 per hour. On the simplest of assumptions, airborne insects would be concentrated by such convergence at an initial rate of 29% increase in area-density per hour. The figure also shows the boundaries of some of the 12 Groups and 103 Blocks which are the operational units of the Sudan Gezira Board, whose 800000 ha of intensive co-operative irrigated agriculture represent one of the largest agricultural enterprises under one management anywhere in the world.

Darfur are so far unsurpassed in terms of all recognized criteria (herd performance, cost effectiveness, net foreign exchange product)' (Adams 1975).

The seasonal migrations of locusts have also been found to be dominated by the i.t.c.z. and other zones of convergence (Rainey 1951), by reason of the down-wind displacements shown by the flying swarms of these insects. Low-level wind convergence moreover will concentrate any airborne insects constrained, e.g. by air temperature, against unlimited ascent and otherwise without systematic movement of their own relative to the air, and this process can have important effects on the population-density of insect pests (Rainey 1976). Sharply defined zones of convergence like the i.t.f. in the Sudan at this season have been found to provide radar line-echoes attributable to the insects concentrated within them (Schaefer 1976).

Rain in the Empty Quarter of Arabia

A further example of widespread and heavy rains in the desert relates to the Rub al Khali, the Empty Quarter of southeastern Arabia, and was investigated in the course of a study (Rainey 1965) of the upsurge of the desert locust plague which took place in the winter of 1948–9. What had been regarded as the last records of the preceding plague had been of swarming locusts which had disappeared both from northeastern Africa and from Indo–Pakistan during the summer of 1948. The first locust reports of what was regarded at the time as a new plague came from the Mudafin and Wadi al Ain areas of the interior of Oman in February 1949, and were provided by the explorer Wilfred Thesiger in the course of a series of famous camel journeys (Thesiger 1950–51) during which he had also encountered exceptional widespread and heavy rains which extended across the eastern sands of the Rub al Khali almost to the Hadhramaut. At Abu Dhabi in late October 1948 (long before the oil finds) Thesiger was 'delayed by a torrential thunderstorm, which flooded a large part of the island and rendered the salt flats impassable to camels', and, en route to Buraimi on 1–4 November, noted that 'the rains had formed extensive lakes on the flats among the dunes, and promised bountiful grazing in the months to come'; the rains had been experienced over an area extending from Buraimi to Sabkhat Mutti, 400 km away to the west.

Across the southern Rub al Khali the information collected by Thesiger shows that these rains also extended eastwards from Wadi Qinab for some 700 km to beyond Mughshin, where flood damage occurred. This was before the time of the use of current weather maps in locust control; and, so far as the Desert Locust Survey of those days was concerned, our first enlightenment as to the reason for these rains was provided by a Pakistani forecaster who chanced to be on night duty at Karachi Airport a year or so later when I called in at the meteorological office there for a casual look at their synoptic charts while awaiting a connecting flight. He immediately recalled an exceptional tropical cyclone with an unusual track which had taken it into southeastern Arabia. Figure 3 shows the successive positions of the centre of this cyclone, as deduced at the time in the Aden meteorological office from the normal routine meteorological observations made in the region. This track had been presented in a report (Bradbury 1949) prepared entirely without knowledge of the information collected by Thesiger on the extent of the rains of late October in the interior, which is clearly consistent (figure 3) with the inland track of the cyclone as inferred from the synoptic evidence.

The weather experienced along the track of the cyclone is illustrated by observations made at the aerodrome at Salala, beginning on the 23rd, two days and some 750 km ahead of the passage of the centre, with an unusual wind of 45 km/h from the north at 16 h (local time), a time of day at which a southeasterly sea-breeze may usually be expected. Distant lightning was noted at 22 h that evening, and the following morning (24th) the northerly wind, still blowing at 35–55 km/h, brought rising sand and reduced visibility to 2 km. At 11 h intermittent drizzle began, developing by later afternoon into intermittent moderate rain, falling through a persistent thick haze which obscured the sky, and becoming heavy though still intermittent in the late evening. The following morning (25th), the persistent northerly wind, of near gale force

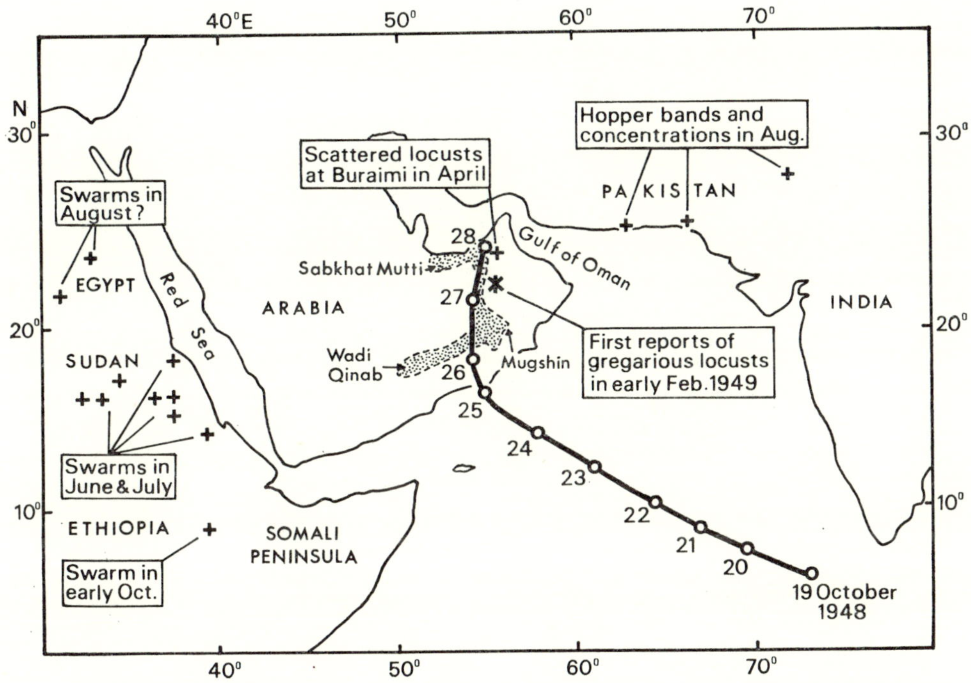

FIGURE 3. The 1948 Oman cyclone: rains, track and locusts. Extent of rain in Arabia as established by information from bedouin (Thesiger 1950–51); track of cyclone established independently by synoptic analysis at Aden meteorological office (Bradbury 1949); and reports of desert locust populations in 1948 which may have been subsequently affected by the cyclone.

(55 km/h at 10 h), which had already caused damage to local shipping and airfield buildings, abruptly died down at 11h15, though drizzle continued: at 11h20 the wind veered to the south, suddenly freshened again, 'began to blow corrugated iron sheets about the camp as if they were autumn leaves', and was recorded as southwesterly 45 km/h at 13 h, 55 km/h at 16 h and 80 km/h (strong gale) at 19 h. On the morning of the 26th the wind had abated, and there were breaks in the cloud cover after 10 h, but further rain and drizzle fell during the day and until 05 h on the morning of the 27th, making a total of 157 mm recorded over the four days, comprising successive daily totals of 48, 41, 43 and 24 mm: the normal total *annual* rainfall at Salala, averaged over a 20 year period (1946–65), is 112 mm. In the hinterland the information collected by Thesiger shows that the rains extended over a WSW/ENE distance of some 700 km. At Masira, 640 km to the northeast away from Salala and from the track of the centre of the cyclone, there was heavy rain all night on the 24/25th, totalling 19 mm and accompanied by strong easterly winds, still 45 km/h from ENE at 07 h on the 25th, but recorded as veering to SSE at 08 h as the rain stopped. At midday on the 26th, rain was reported by an

aircraft over the Gulf of Oman, extending for 400 km along 25° N, while at Sharjah, on the Trucial coast of the Persian Gulf, 19 mm were recorded on the 25/26th, and rain was again reported on the 28th.

Figure 4 shows that the area over which surface winds were affected by the cyclone extended from the Gulf of Oman to beyond the southern Red Sea, where both at Assab in Eritrea and at Perim the normally persistent southeasterly winds characteristic of this season of the year were

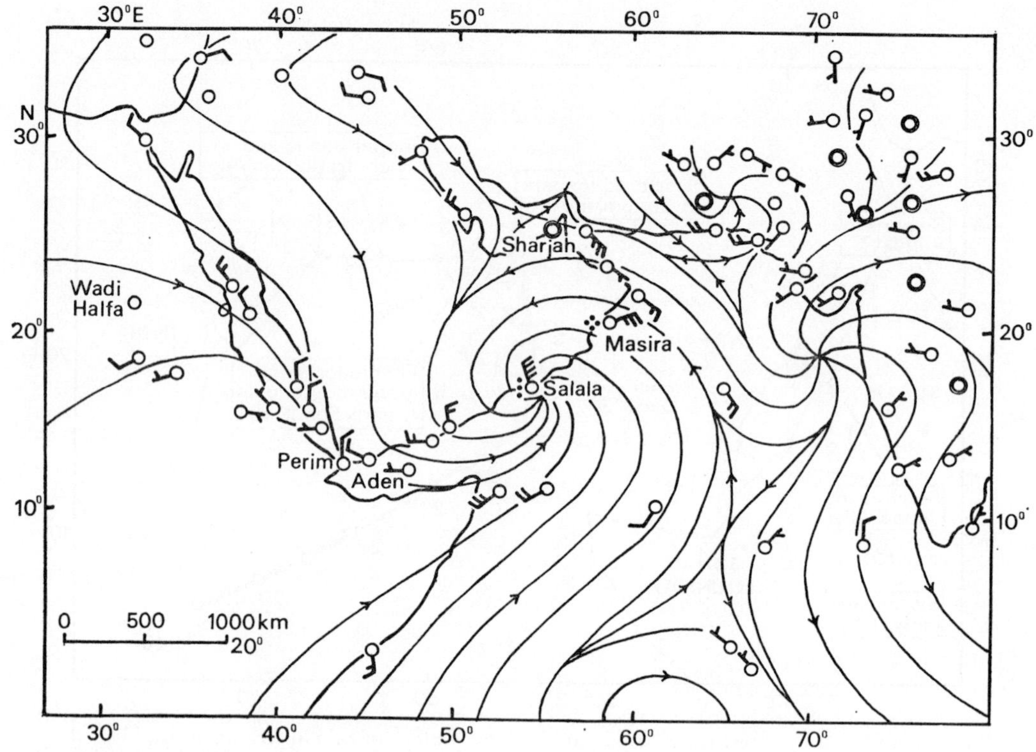

FIGURE 4. The Oman cyclone: surface winds and streamlines at 03 h G.M.T. 25 October 1948.

interrupted by an exceptional two-day spell of northerlies and northeasterlies on 25 and 26 October. In view of the down-wind direction of displacement which has been found so consistently in airborne locust populations, this meteorological evidence thus supported a suggestion that locusts derived from swarming populations in the Red Sea and/or in Baluchistan might well have participated in breeding in the eastern Rub al Khali in November 1948, in addition to any remaining local solitary-living locust populations, and that the widespread rain would have provided extensive conditions suitable for the multiplication of all locusts involved, with a probable rate of development appropriate for the following generation to reach the stages reported by Thesiger in the following February.

RAIN AROUND THE KALAHARI

The strongly convergent winds of the Arabian cyclone are also likely to have concentrated any locusts in flight in them, though the observations in this case were inadequate for any quantitative estimate of possible rates of concentration. Because of the importance attached to possible changes in the population density of the locusts in the development of gregarious behaviour, however, attention had been directed to the corresponding findings of a series of studies made at about this same time on the dynamics of a better-documented and somewhat, analogous cyclone from the southern Indian Ocean (Black 1950; Kreft 1953; Sellick 1958). This cyclone (figure 5) now has relevance for the present meeting by reason of having brought widespread rains to the semi-arid regions bordering on the Kalahari desert and discussed by

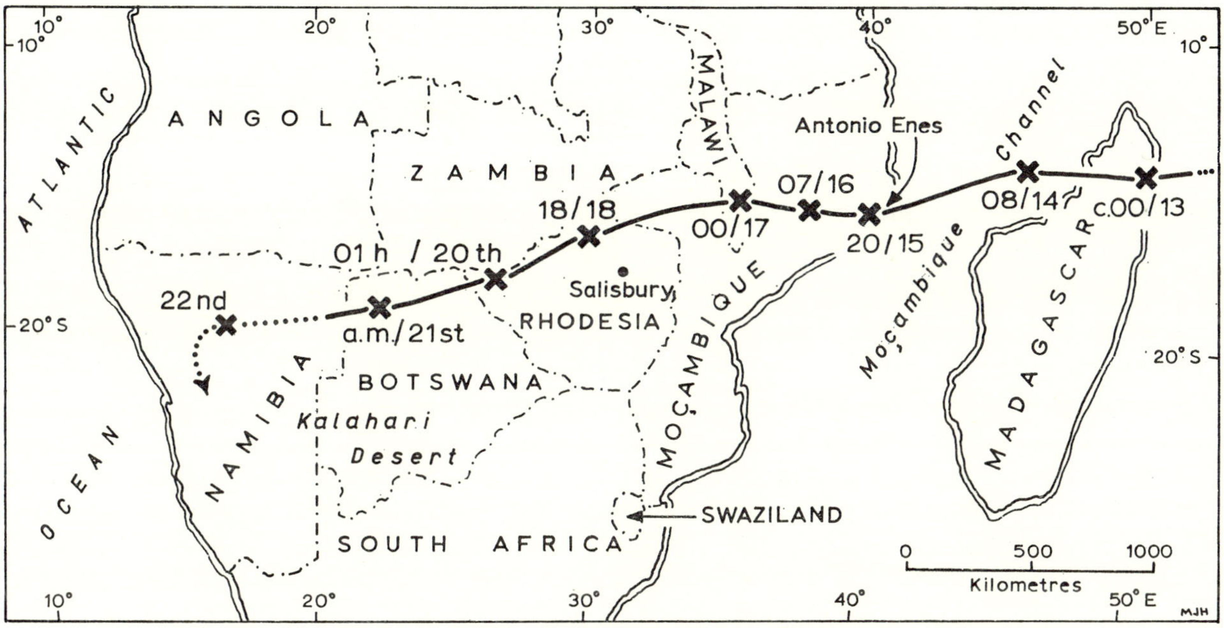

FIGURE 5. Track of an unusual cyclone from the southern Indian Ocean: February 1950. (Data of Black 1950, Kreft 1953 and Sellick 1958.)

Pereira (pp. 555–562). It crossed Madagascar and the Mozambique Channel and then moved inland across Mozambique, Malawi, Rhodesia and Zambia between 16 and 20 February 1950, at a latitude (15°–18° S – instead of N) and a speed (13–14 km/h) closely comparable with the Arabian cyclone just described, and also giving a roughly comparable rainfall: a total of 145 mm fell in three days (13–16th) at Antonio Enes on the Mozambique coast over the period of the passage of the cyclone, followed by a well-documented average daily fall of 52 mm over an inland area of 155000 km^2 on the 18th and 19th, with indications that precipitation of this order had persisted for four or more days. The 1950 cyclone moved on across Botswana on 21 February and was still giving good rains over northern Namibia on the 22nd. There was considerable similarity between the wet-bulb temperatures which had been recorded at Salala, which remained between 22 and 23 °C for at least 12 h during the passage of the Arabian cyclone, and the wet-bulb potential temperatures, between 21 and 22 °C from the ground up to at least 7500 m above sea level, recorded at Salisbury in Rhodesia during the radiosonde ascent on 18 February 1950, which was regarded as representative of the inland rain-area.

It was shown that the total water-vapour content of the atmosphere at the time, as indicated by this radiosonde ascent, was equivalent to only 36 mm of liquid water; and that evaporation and transpiration were not likely to have returned more than about 5 mm of the rainfall per day to the atmosphere. In order to maintain the observed rate of precipitation, the wind-system must therefore have been such as completely to replace, once in every 20 h or so, the whole of the air over the rain area, up to a great height; and this was shown to require a net inflow component of the wind, near the ground, of the order of 7 km/h around the perimeter of an area 440 km in diameter, representing the central part of this wind-system. Hypothetical airborne insects experiencing such convergence would be concentrated at a rate which on the simplest of assumptions would increase their area-density by an average factor of five over an area of more than 10^5 km^2 within a period of four days.

Rain in the outback

The Oman cyclone of October 1948 provided months of bountiful grazing to the bedouin of the Empty Quarter of Arabia. A somewhat similar meteorological system which benefited the cattlemen of the Australian outback in the same way was cyclone 'Audrey' of 1964 (figures 6 and 7), of which we heard a great deal in the course of field work around Nockatunga in this area the following year, under the auspices of the C.S.I.R.O. Division of Entomology (Clark, Ashall, Waloff & Chinnick 1969). I am greatly indebted to Derek Reid, of the C.S.I.R.O. Division of Meteorological Physics, for these analyses, and to the Director of the Commonwealth Bureau of Meteorology for the basic data and routine charts.

Rainfall amounts of more than 200 mm were recorded in the vicinity of the track of 'Audrey', with the fall over a period of a week in a number of places approaching the median annual rainfall (Gibbs 1969) for the area. Local systems for the exploitation of such rains, epitomized in the term 'opportunity country' (Ricketts 1969), were illustrated by an extensive cattle station on which we worked, in an area with a median annual rainfall of about 200 mm and carrying relatively few cattle in drier years. This station was owned and operated, in the way indicated by Tribe (p. 565), by a syndicate which also owned a second station in the higher-rainfall Mitchell River area some 1300 km away to the north with a median annual rainfall of more than 1000 mm. From this second station, after rains like those of 'Audrey', store cattle were brought in for fattening on the resulting extensive but temporary grazing, using 'beef-trains' of heavy road trucks with multiple trailers for the rapid transport of stock without loss of condition.

Cyclone 'Audrey' had incidentally been followed by an upsurge of the Australian plague locust (*Chortoicetes terminifera* Walk.), though the evidence available was insufficient to substantiate transport and concentration of flying locusts by the convergent wind-system of the cyclone in a manner suggested by analogy with the Oman cyclone of 1948. However, the late Dr D. P. Clark, of the C.S.I.R.O. Division of Entomology, subsequently directed attention to the fuller and acceptable evidence on this same point which was secured by his field staff (particularly R. A. H. Davies) on the locust developments in 1971–72 which were associated with the passage of another cyclone across the same general area.

Rainfall in such areas does not have to be on this scale to be associated with recognizable synoptic systems. Thus for example during our own period of field work in and around south-western Queensland, in January–March 1965, rains were less than 50 % of normal; but each of the four occasions during this period on which our party noted rain was subsequently found to have been associated with the passage of a trough of low pressure, shown at the time on the routine synoptic charts of the Commonwealth Bureau of Meteorology. These features were marked locally by extensive cirrus and alto-cumulus cloud and by freshening winds shifting

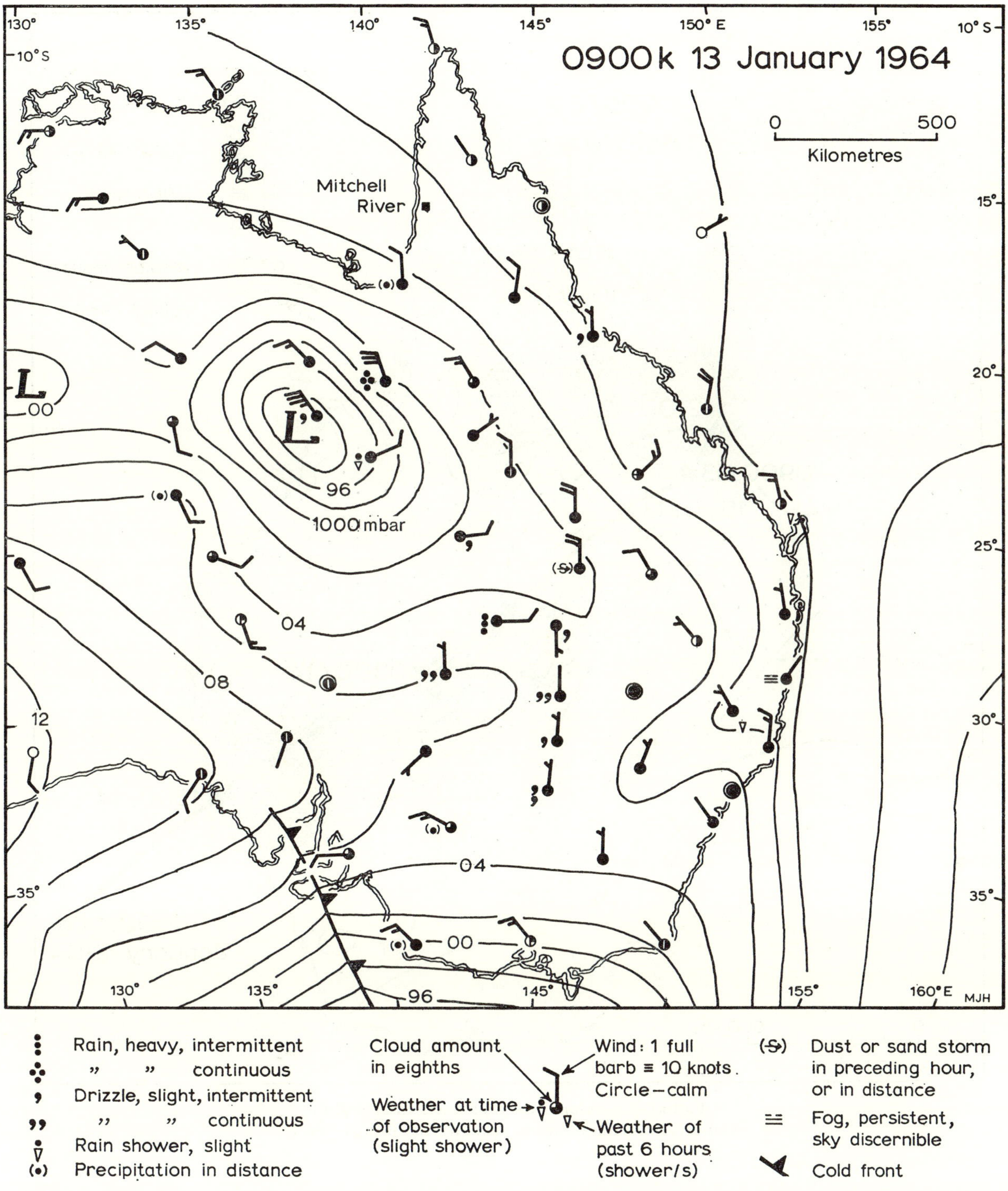

FIGURE 6. Cyclone Audrey: surface synoptic analysis.
(Data and analysis of Commonwealth Bureau of Meteorology.)

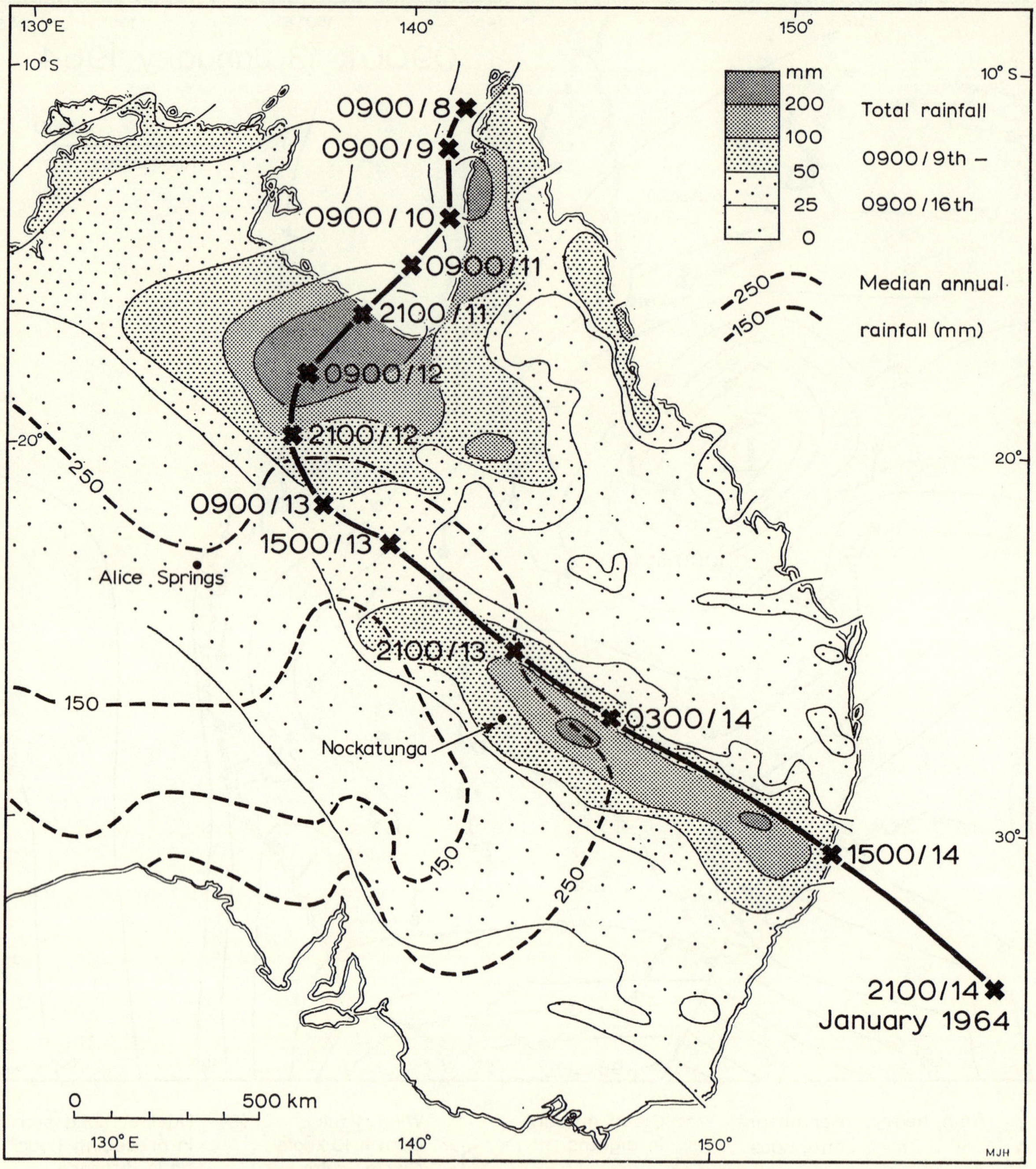

FIGURE 7. Cyclone Audrey: track and rainfall.
Isohyets of median annual rainfall from Gibbs (1969).

from the usual easterlies and southeasterlies into the northerly and westerly quarters, as well as by records of rain at one to nine of the 21 regular rainfall stations within 300 km of Nockatunga. The distinction shown by these data, between extended spells of completely dry weather and shorter unsettled spells with rain recorded simultaneously at several points, was demonstrated even more strikingly by the records of six stations within 100 km of Alice Springs in central Australia during the drought year of 1961: the 69 occasions during this complete year on which rain was recorded at one of these six stations were confined to 36 of 365 possible days, occurring in 23 sequences of up to four successive days with rain, and separated by completely dry spells of up to four months (Newsome 1965).

Rain in the southwestern Sahara

A final example comes from West Africa: on 9–10 February 1968 widespread rains fell in the western Sahara, over an area whose extent is directly illustrated by the range of the stock movements undertaken by the nomads of the region in exploiting the resulting grazing. For details of this particular occasion, and for much other information on this subject ('le Sahara n'est pas du tout une région abiotique'), I have been very greatly indebted to M. Abdallahi Ould Mohamed Sidia, Director-General of O.C.L.A.L.A.V. (Organisation Commune de Lutte Antiacridienne et de Lutte Antaviare), the international organization responsible for the control of desert locusts and other migrant pests in nine countries of western Africa, from Mauritania and Senegal to Chad and Cameroun. M. Abdallahi comes himself from Mauritania, from a community of nomads who still make regular seasonal migrations; his own recollections go back to the rains in northern Mauritania in February 1935, flooding the clay plain of Rhallamane and recalled as the 'year of the truffles', from the truffle *Terfezia* which grows on the desert rock-rose *Helianthemum.* These 1935 rains were experienced and recorded by Monod; the scope and value of the collection and critical analysis of such information from the nomads on the rainfall régime and ecology of the Sahara has been recognized in many studies, such as those of Dubief on rainfall and run-off.

The rains recorded during 9–10 February 1968 reached 41 mm, at Akjoujt; and figure 8 shows two noteworthy stock movements which were made possible by the resulting grazing; a fall of 20 mm has been found sufficient to initiate the growth of useful grazing (Toupet 1975). In one of these movements, cattle were brought from the Moudjeria area, the northern limit of their usual distribution, over a distance of some 500 km to within 60 km of Bir Moghrein – a feat providing evidence of the degree of continuity of the vegetal cover and in turn of the rains which had evoked it. The second was a movement of other stock between the Bir Moghrein area and In Nabnane, 1000 km away on the Mali/Algeria border – a movement which is understood to have exploited not only the grazing but also certain gaps in the customs coverage of these frontiers. Clearly these rains must have covered a total area of hundreds of thousands of square kilometres, at a time of year when losses by evaporation are much less than they would be in summer (cf. Perry, p. 493 – on differences between summer and winter rainfall cropping limits around the Sahara); and the value of the resulting grazing is not merely that of its extent. The high nutritive value of such temporary winter grazing in the Sahara has helped to make it the traditional complement to the summer grazing of the Sahel to the south; and the very great value of such stock movements between complementary seasonal grazing areas, in resting land at present heavily overgrazed, is now well recognized (Monod 1975, etc.).

In the exploitation of the grazing following such desert rains the Sahara nomads use not only a very extensive information network of their own but also an impressive series of traditional techniques, ranging for example from the estimation of the distance of a rainstorm from observations of the associated lightning, to the use of the desert crucifer *Schouwia* for promoting fertility in livestock; and their vocabularies include for example different words for the characteristically convective showers of afternoon and evening and for the more general and less stormy rains of night and morning.

Contemporary technology provided information on these rains of February 1968 illustrated both by satellite photography on cloud coverage (shown in figure 8 as the coded nephanalysis which was received in Kensington a few hours later, as part of the desert locust forecasting system of those days), and by the evidence on the mechanism of these particular rains which was

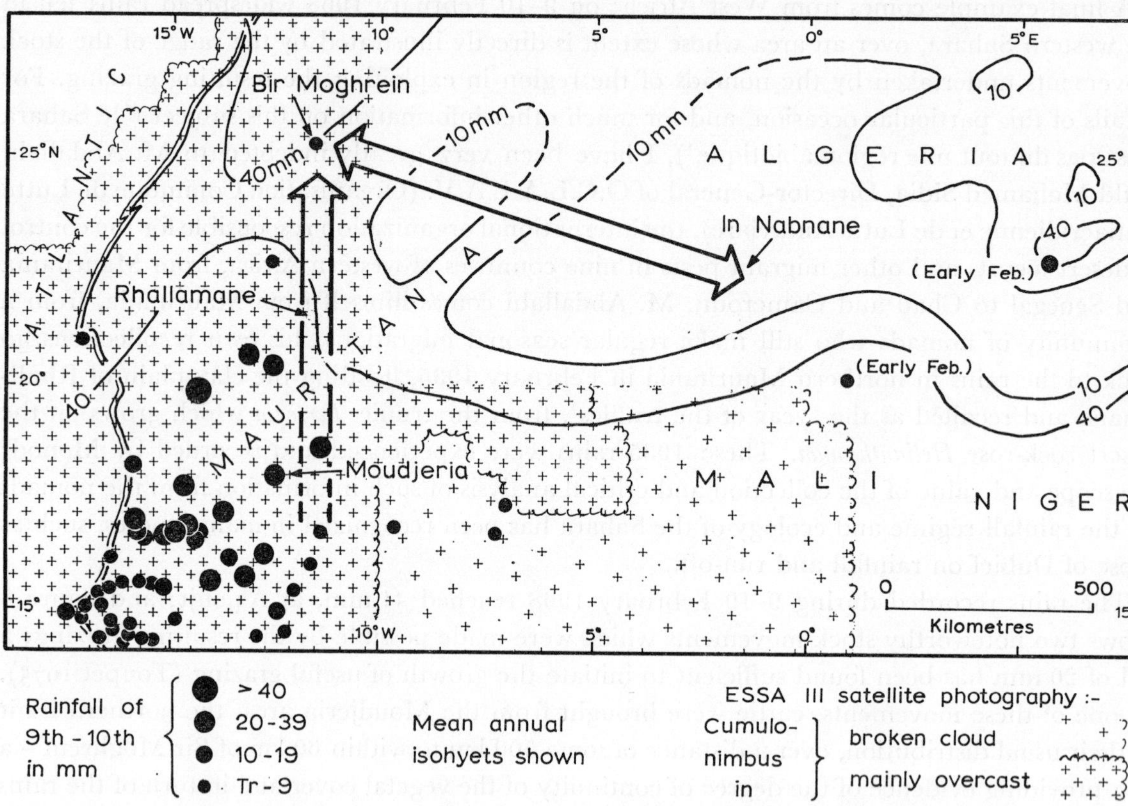

FIGURE 8. Rains in western Sahara, February 1968: exceptional stock movements on subsequent grazing. (Rainfall observations from A.S.E.C.N.A. Dakar; annual isohyets from Dubief (1959–63).)

supplied by current synoptic analyses. The rains were associated with the passage of a higher-latitude disturbance which was dominating the weather as far north as Britain, and on which I am greatly indebted to M. Robert Garnier, of the Dakar Bureau d'Etudes of A.S.E.C.N.A. (Agence pour la Securité de la Navigation Aerienne en Afrique et à Madagascar) for the benefit of his synoptic experience and for the following comments:

'Typical interaction between meteorological phenomena of temperate and tropical latitudes: surge of polar front brought substantial invasion of cold air to low latitudes over Atlantic. At 1200 GMT 9 February this deepened a trough of low pressure which extended at all altitudes from a depression over the British Isles to Cape Verde, and the gradient between this trough and high pressures over the Sahara produced a strong SW current (exceeding 220 km/h at heights from 7500 up to 16000 m between Dakar and Nouadhibou) which brought up equatorial air giving the cloud, thunderstorms and rains over Senegal and Mauritania. This situation began on the evening of the 8th and lasted until the night of 10/11th.'

Conclusion

It is suggested that the full exploitation of the considerable resources represented by rains such as those described – as well as the timely and economic control of upsurges of migrant pests themselves exploiting such rains – requires the integration of the information networks of international meteorology with those of the desert nomads; and attention is directed to this opportunity and challenge of bringing to bear the modern observational resources and the synoptic expertise of contemporary meteorology on the basic human needs not only of the desert nomads, but also of their neighbours in a future of increasing pressure on global food supplies. More specifically, there already is an internationally supported organization, with terms of reference very reasonably appropriate for the provision of such an integrated information service on current rainfall and grazing conditions, and very recently established in a highly appropriate area, in the shape of Dr Rijks' U.N.D.P./F.A.O. Sahel Centre for Agrometeorology and Applied Hydrometeorology at Niamey in the Niger Republic.

References (Rainey)

Adams, M. 1975 Savanna development project (Phase II): working paper no. 3. Mimeogr.

Black, A. A. 1950 Extraordinary track of a cyclone. *Weather, Lond.* **5**, 229–230.

Bradbury, T. A. M. 1949 An account of a tropical cyclone in the Arabian Sea, October 1948. Meteorological Office, Aden (unpublished).

Clark, D. P., Ashall, C., Waloff, Z. & Chinnick, L. 1969 Field studies on the Australian Plague Locust in the Channel Country of Queensland. *Anti-Locust Bull.* **44**, 1–101.

Dubief, J. 1959–63 *Le climat du Sahara.* Algiers: Université d'Alger.

Gibbs, W. J. 1969 Meteorology and climatology. In Slatyer & Perry (1969), pp. 33–54.

Hadley, G. 1735 Concerning the course of the general trade winds. *Phil. Trans. R. Soc. Lond.* **29**, 58–62.

Kreft, J. 1953 Descriptive account of a cyclone of February 1950. *Notos, Pretoria* **2**, 149–156.

Magor, J. I. 1962 Rainfall as a factor in the geographical distribution of Desert Locust breeding areas. Ph.D. Thesis, Edinburgh University.

Mason, B. J. 1976 Towards the understanding and prediction of climatic variations. *Q. Jl R. met. Soc.* **102**, 473–498.

Monod, T. (ed.) 1975 *Pastoralism in tropical Africa.* Oxford: International African Institute.

Newsome, A. E. 1965 Data in Clark *et al.* (1969).

Osman, O. E. & Hastenrath, S. L. 1969 On the synoptic climatology of summer rainfall over central Sudan. *Arch. Met. Geophys. Bioklim.* (B) **17**, 297–324.

Pedgley, D. E. 1974 An exceptional desert rainstorm at Kufra, Libya. *Weather, Lond.* **29**, 64–71.

Rainey, R. C. 1951 Weather and the movements of locust swarms: a new hypothesis. *Nature, Lond.* **168**, 1057–1060.

Rainey, R. C. 1965 Note on a tropical cyclone affecting southern Arabia in 1948. *World Met. Org. Tech. Note* **69**, 257–260.

Rainey, R. C. 1976 Flight behaviour and features of the atmospheric environment. *Symp. R. entomol. Soc.* **7**, 75–112.

Ricketts, W. J. 1969 The role of transport in the utilisation of arid land resources. In Slatyer & Perry (1969), pp. 142–157.

Schaefer, G. W. 1976 Radar observations of insect flight. *Symp. R. entomol. Soc.* **7**, 157–197.

Sellick, N. P. 1958 Modification of the structure of a tropical cyclone on traversing a high plateau. *Q. Jl R. met. Soc.* **84**, 259–269.

Slatyer, R. O. & Perry, R. A. (eds.) 1969 *Arid lands of Australia.* Canberra: A.N.U. Press.

Thesiger, W. 1950–51 Desert borderlands of Oman. *Geogr. J.* **116–117**, 137–171, 128.

Toupet, C. 1975 Le nomade, conservateur de la nature? L'example de la Mauritanie centrale. In Monod (1975), pp. 455–467.

World Meteorological Organization 1967 *Climatological normals for the period* 1931–1960. Geneva: W.M.O.

Discussion

E. G. Davy (*World Meteorological Organization, Case Postale No.* 5, *CH*–1211–*Geneva*). It is not very uncommon in the Northern Sahel, or even in the Sahara, for the occasional period of very heavy rain in an arid area to be followed by abnormally good growth of animal grazing to which the nomadic herds are attracted from great distances as in the case cited for northern Mauritania.

I would sound a note of warning on the attractive suggestion that the existing communications systems of the nomads should be perfected by modern technologies, including monitoring by earth satellites, to make greatest use of such grazing opportunities.

Such localities provide the seed banks which are replenished perhaps once in five or ten years in arid regions. From these, much wider areas are seeded for future years by wind dispersion. Larger numbers of animals directed to those areas would not only seriously diminish seed production but also remove the vegetation cover of the soil to expose to increased desertification those areas which are already at greatest risk.

R. C. Rainey. I fully accept Mr Davy's note of warning, which is, however, particularly relevant to the overgrazed areas of the northern Sahel in and around which stock are almost continuously present. The important point about the kind of rains which I have been describing in the Sahara, further to the north, is their extent, which may amount to hundreds of thousands of square kilometres on a single occasion, and over which the effects of additional stock movements, like those recorded on p. 451, might be expected to be small, while at the same time helping to relieve grazing pressure in the Sahel.

G. Stanhill (*Volcani Center of Agricultural Research, Israel; for correspondence, Botany Dept, Cambridge University*). A key to the successful exploitation of scarce rainfall is the recognition of its large interannual variability. Once this is acknowledged, settled agriculture appears possible in surprisingly low rainfall areas. For example, in southern Israel, attempts to crop annually with winter cereals an area with an average annual rainfall of 200 mm and a coefficient of variation of 35 % had to be abandoned when a run of wetter-than-average years came to an end. Experiments to replace cereal cropping with a flexible system of sheep farming on perennial pastures appear to be successful and show promise of long term stability.

Further south, Professor Evanari's reconstruction of Nabatean farms at Avdat has shown that it is possible to crop perennially deciduous fruit orchards with a mean annual rainfall of less than 100 mm by exploiting the special runoff and infiltration properties of the loess soil.

E. B. Worthington (*I.B.P., c/o The Linnean Society, Burlington House, Piccadilly, London W*1*V* 0*LQ*). The question of the distribution of pests and vectors of disease through climatological disturbance such as thermals lifting small organisms from the water or land for subsequent movement by prevalent winds: were there examples of this other than the well established movements of locusts? Climatological disturbance might for example account for the recent somewhat unexpected reappearance of adult *Simulium damnosum,* infected with onchocerciasis, in parts of the zone in West Africa from which this vector had already been eliminated in the great experiment being undertaken by W.H.O. which is designed to eradicate river blindness from a huge area in West Africa.

R. C. Rainey. Other pest organisms subject to important redistribution by air movements include the spores of rust-fungi. Thus 'an annual double transcontinental migration through the atmosphere is an essential condition for the development of the rust epidemics [of *Puccinia graminis* and *P. triticina*] which regularly attack cereal crops in North America' (Gregory 1961). Again, the progress of the 5-year spread of American maize rust (*P. polysora*) across Africa following its first arrival, in Sierra Leone in 1949, showed a remarkable analogy with the 5-year spread of the swarms of the last major outbreak of the migratory locust (*Locusta migratoria migratorioides*) across Africa following their first appearance, in Mali in 1928 (Rainey 1973).

References

Gregory, P. H. 1961 *The microbiology of the atmosphere.* London: Leonard Hill.
Rainey, R. C. 1973 Airborne pests and the atmospheric environment. *Weather, Lond.* **28**, 224–239.

D. E. Pedgley (*Centre for Overseas Pest Research, College House, Wrights Lane, London W8 5SJ*). At the request of the World Health Organization, the Centre for Overseas Pest Research had a first look in 1975 at the evidence for downwind movement of *Simulium damnosum*, the vector of river blindness in West Africa. Both the few published accounts and some current season's records from the W.H.O. control programme were used, with the tentative conclusion that first appearances of biting flies after the dry season were associated with convergence zones in the wind-field – both the intertropical convergence zone and the smaller-scale zones in squall lines. Much work needs to be done, not only on tracking flies in changing wind fields but also on the flight behaviour of individual flies, including their height, endurance and timing of flight.

R. C. Rainey. A brief spell of field work for the W.H.O. Onchocerciasis Control Programme, undertaken in Upper Volta and Ivory Coast during August 1976 which used the same Doppler-equipped aircraft mentioned on p. 441, is already helping to throw light on the wind-systems which appear to be involved in the crucial problem of re-invasion to which Dr Worthington has referred.

W. B. Broughton (*Animal Acoustics Unit, City of London Polytechnic, Calcutta House, Old Castle St, London E1 7NT*). Dr Rainey spoke of the air drying out as it rises in the tropical zone. What happens to the water it loses, and, when a disturbance of the usual atmospheric circulation brings the infrequent rain to an area of convergence, where does it pick up this water?

R. C. Rainey. The water lost by the air rising in the tropical zone provides the rains of this zone (p. 440), and the water which provides the rain associated with convergence in an atmospheric disturbance in a normally arid region will, likewise, have been picked up from the tropical oceans.

Phil. Trans. R. Soc. Lond. B. **278**, 457–475 (1977) [457]
Printed in Great Britain

The geography of semi-arid lands

By A. T. Grove
Department of Geography, Cambridge

[Plate 1]

Semi-arid lands were at times during the last 20000 years more humid and at other times more arid than at present, with important consequences for the soil and water resources of the present day. These lands were the scene of the beginnings of pastoralism, the cultivation of cereals, and urban living between 10000 and 5000 years ago. The environment has always been attractive to man but it is liable to deteriorate towards desert with long-lasting consequences. Development possibilities diversify according to environmental conditions which vary from place to place and also according to the economic and other characteristics of the States in which these lands are situated. The States can be regarded as falling into four categories: there are on the one hand wealthy countries – either industrialized or rich in minerals; on the other hand there are poor countries – some with land outside the arid zone as yet not fully developed, others with no land of this kind. In response to the differences in economic opportunity, migratory movements are taking place, notably from the impoverished semi-arid regions to the towns, particularly those in the petroleum-producing states.

1. Semi-arid climates

The semi-arid lands are those parts of the world where the rain is insufficient or barely sufficient for satisfactory crop growth in most years. Such regions can be defined only in an arbitrary manner. A map which was prepared by Meigs (1953) for UNESCO has been widely accepted as a useful guide, for example by Hills (1966) in a book bringing together some of the results of UNESCO's Arid Zone Programme. McGinnies, Goldman & Paylore (1968) used maps of the continents, in *Deserts of the world*, depicting the arid zone in much the same way as Meigs had done, and similar maps appear in *Arid lands in transition* (Dregne 1970) and also in Dov Nir's *Semi-arid lands* (1974). Based on Thornethwaite's (1948) aridity coefficient, the UNESCO map is derived from the relation between precipitation, potential evapotranspiration, water surplus and water replenishment, calculated for individual months throughout the year. Indices distinguish the semi-arid from the arid lands, and extremely arid lands are defined as those where 12 successive months have been recorded without rain.

The extremely arid lands correspond roughly with those having mean annual rainfall totals of less than 25 mm. Arid lands between the tropics correspond with those receiving between about 25 and 200 mm; semi-arid lands approximate to those areas with mean annual rainfall totals of 200–500 mm. Towards the poles, mean annual precipitation figures corresponding to the limits of arid and semi-arid land diminish as evapotranspiration diminishes (figure 1).

Aridity in low latitudes is associated with the main tropical high pressure systems and with dry air descending on the poleward sides of Hadley cells. Rain falls in the summer season, when the high pressure cells lie poleward of their mean positions and humid equatorial air reaches higher latitudes. Rainfall is typically episodic with local storms a few kilometres in diameter moving across country. The main growing season starts with the soils thoroughly dry and

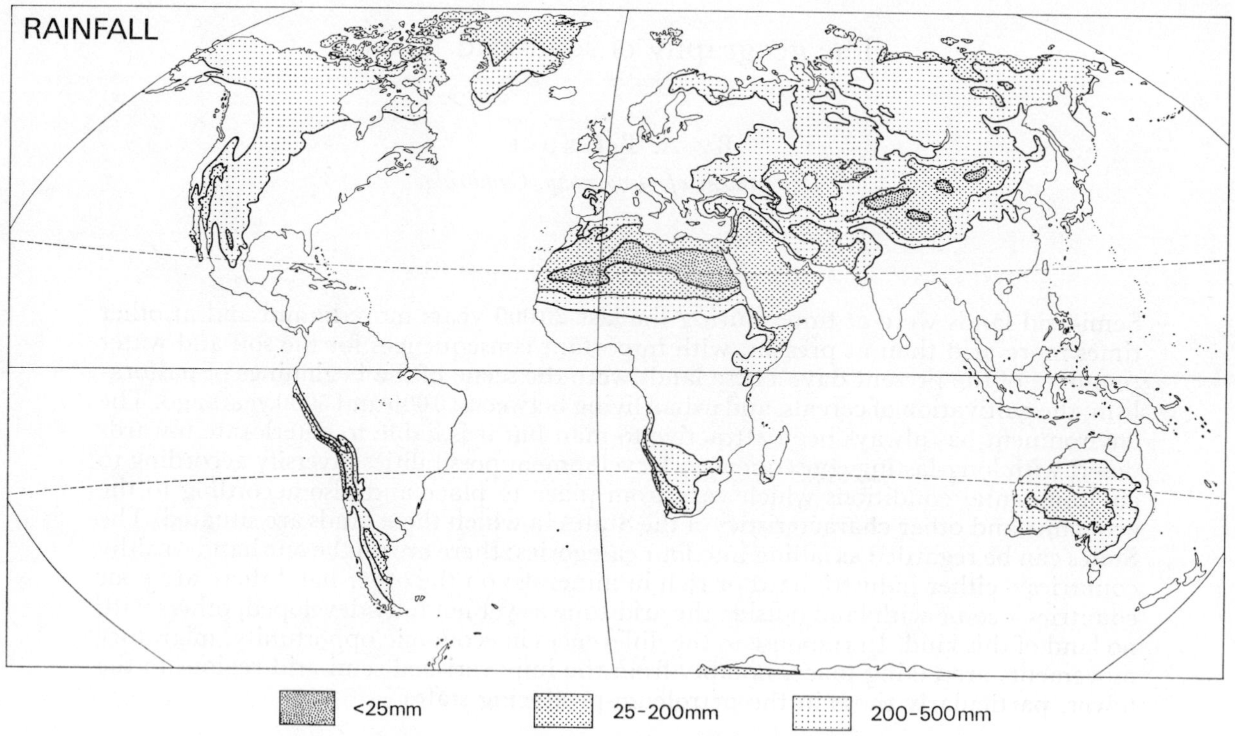

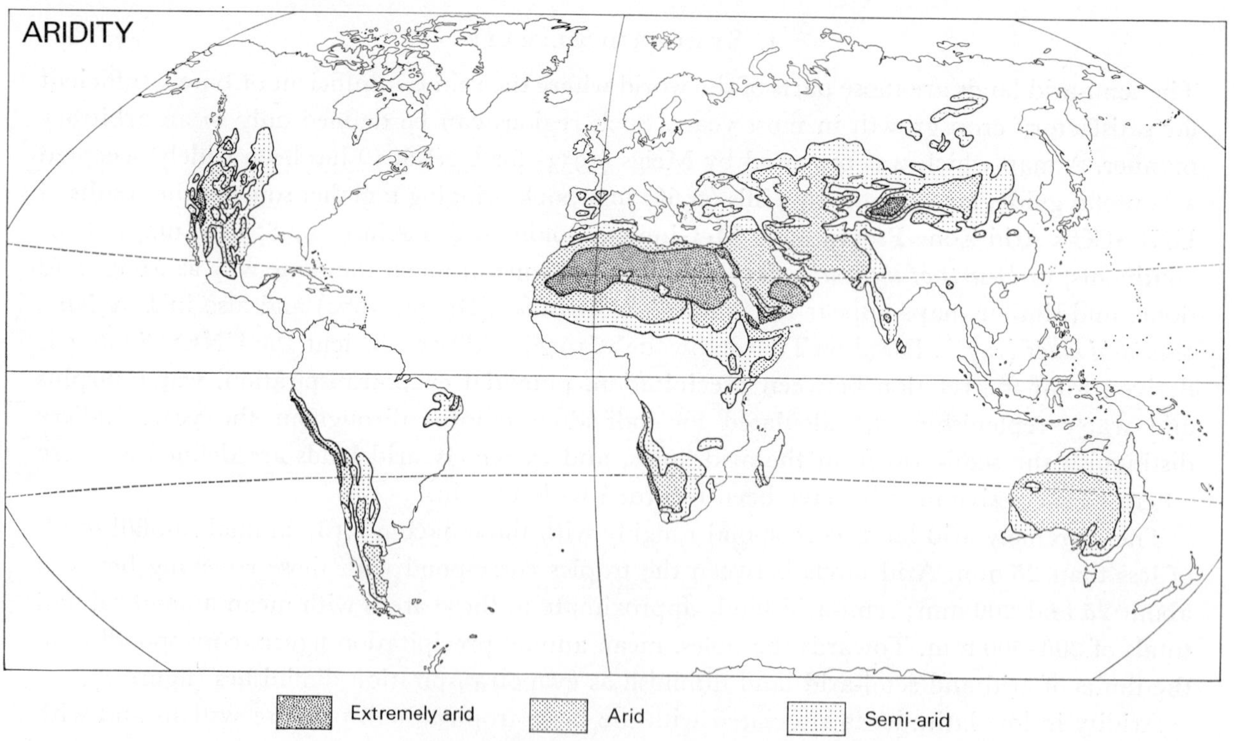

FIGURE 1. (*a*) Mean annual precipitation. (*b*) Extremely arid, arid and semi-arid regions (after Meigs 1953).

consequently risks of crop failure are high if the early rains are interrupted by a dry spell. Such is the case if the poleward movement of the equatorial air is indecisive.

On the poleward sides of the tropical deserts, lands under the influence of the tropical high pressure systems and therefore dry during the summer season receive rain from mid-latitude depressions during the winter months. In the spring, soils are moist at depth and crop growth is, in this respect, less hazardous than in areas with similar mean annual rainfall totals lying between the tropics. In many such areas supplementary irrigation water is available from neighbouring mountain catchments and, in some places, from glaciers.

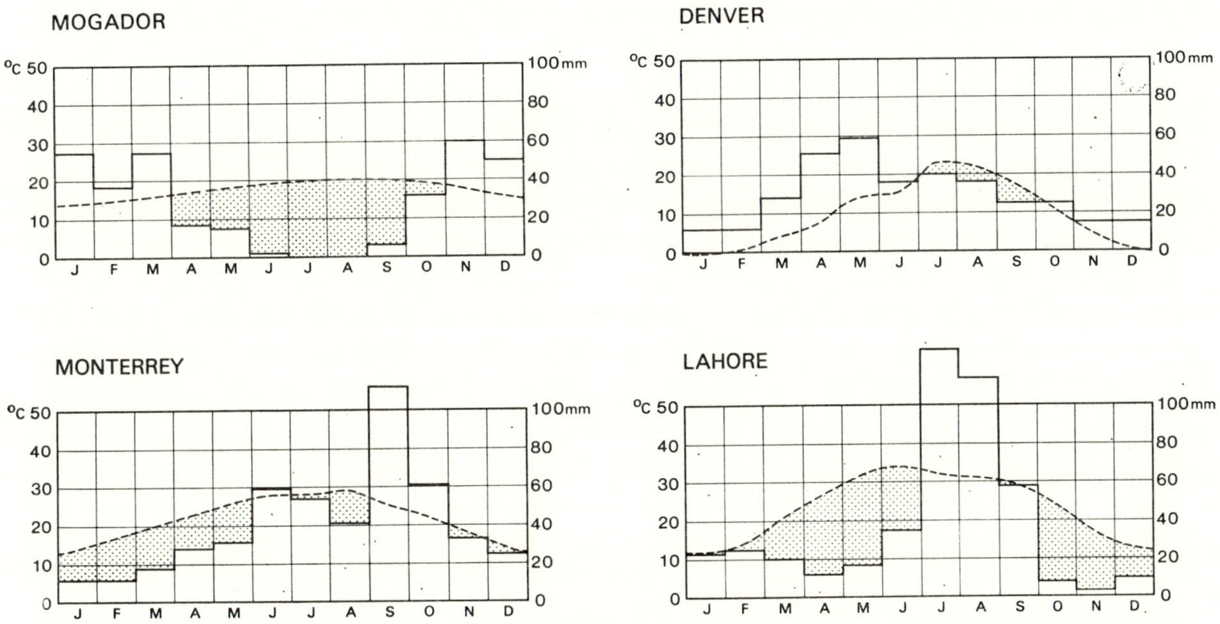

FIGURE 2. Mean monthly precipitation (solid lines) and mean monthly temperature (dashed lines) for Mogador, Morocco; sub-tropical west coast Denver, Colorado, USA; mid-latitude continental Monterrey, Nuevo Leon, Mexico; tropical highland Lahore, Pakistan, tropical continental semi-arid.

Mid-latitude aridity is more closely related to the global distribution of land and sea. The most extensive area of low rainfall extends from the tropical North Atlantic across the Sahara and Arabia into the heart of the Asian landmass. It is bordered to north and south by semi-arid country stretching from Asia Minor and the regions bordering the Caspian Sea, across central Asia into Mongolia and China. Other extensive semi-arid lands in middle latitudes lie on the leeward sides of mountain ranges running across the track of westerlies, in western North America from the Gulf of Mexico to the southwestern Canadian prairies, in Argentina and in eastern Europe (figure 2). Precipitation is more evenly distributed through the year than nearer the equator, but the growing season is limited by low temperatures persisting well into the spring and by autumn frost and snow. Winter snowfall is commonly an important source of moisture for agriculture and for river flow.

Maps showing the distribution of precipitation over the globe show steep gradients of raininess, not only across mountain ranges but over the oceans as well. Consequently quite small shifts in global circulation patterns in the atmosphere or oceans can produce marked changes in the rainfall received at particular stations, especially in the months at the beginning and end of the rainy season. Such variations from year to year and over longer periods of time are well marked over the oceans and at their margins as well as over the continents. Surface water temperatures, which are related to the vigour of upwelling and the strength of the Trade Winds, have a great effect on precipitation (Musk 1976). Records of annual rainfall from oceanic islands in the tropics exhibit persistence of high or low values over periods of several years (Aspin 1976). Continental margins may be involved in such variations and although their climate means may characterize them as semi-arid, conditions vary from intensely arid for years on end to disastrously wet for several months or years.

Individual years and sequences of years with rainfall totals well below median values also occur in continental interiors. In 1972, for example, the zone of semi-arid land south of the Sahara was displaced between 100 and 200 km south of its mean position; the rains in several of the preceding years had also been deficient and the following year was a dry one. The resulting sufferings of people and livestock living in the region attracted world attention. In that same year, 1972, a warm current of equatorial water, El Niño, spread down the west coast of South America (Caviedes 1975). In place of cold, upwelling water, with abundant nutrients and fishlife, warm water reached more than 12° S, and instead of aridity there were heavy rains on the coast. Catches of fish were reduced from about 9 million to about 4 million tonnes. That year rainfall was below median values in the semi-arid lands of Mexico, the U.S.S.R., and India. Some climatologists were inclined to see the decline in rainfall as being associated with a global climatic fluctuation that was likely to be long-lasting. Some suspected that man's activities were in part responsible, particularly the consumption of fossil fuel and pollution of the atmosphere with dust and other substances. However, rainfall records show that runs of years with rainfall below the median are to be expected, and it has not been shown that any long-term displacement of the semi-arid zones is in progress. It has been made clear to all that climate is of its nature variable and that the semi-arid lands with their high evapo-transpiration losses are particularly sensitive to rainfall deficiencies.

2. Effects of Late Quaternary climatic changes

The evidence for climatic fluctuations before the instrumental period, is difficult to interpret. There are indications from long Chinese historical records (Chu Ko Chen 1973) and from elsewhere that shifts in the climatic boundaries of the kind that occurred around 1972 have happened in the past and that increased aridity has persisted over much long periods of time. In general, however, it would seem that we may have experienced in the last century or two the full range of variations in climate that have affected the globe over the last three or four thousand years. Studies that have been made in a number of different fields, especially since 1960, have shown that the positions of the climatic boundaries shifted more violently over the longer time period of the Late Quaternary, in low latitudes as well as in high latitudes. It appears that the mean values of precipitation in tropical Africa moved through a wider range in the Late Quaternary than have extreme values for individual years in this century.

For the globe as a whole, three periods of different levels of biological activity can be distinguished in the last 20000 years. The first period might be called a state of low activity, the second a state of high activity, and the third a state of reduced activity.

At the height of the last glacial episode, from some time before 20000 to about 13000 years ago, global biological activity and terrestrial biomass were at low levels. Ice desert occupied

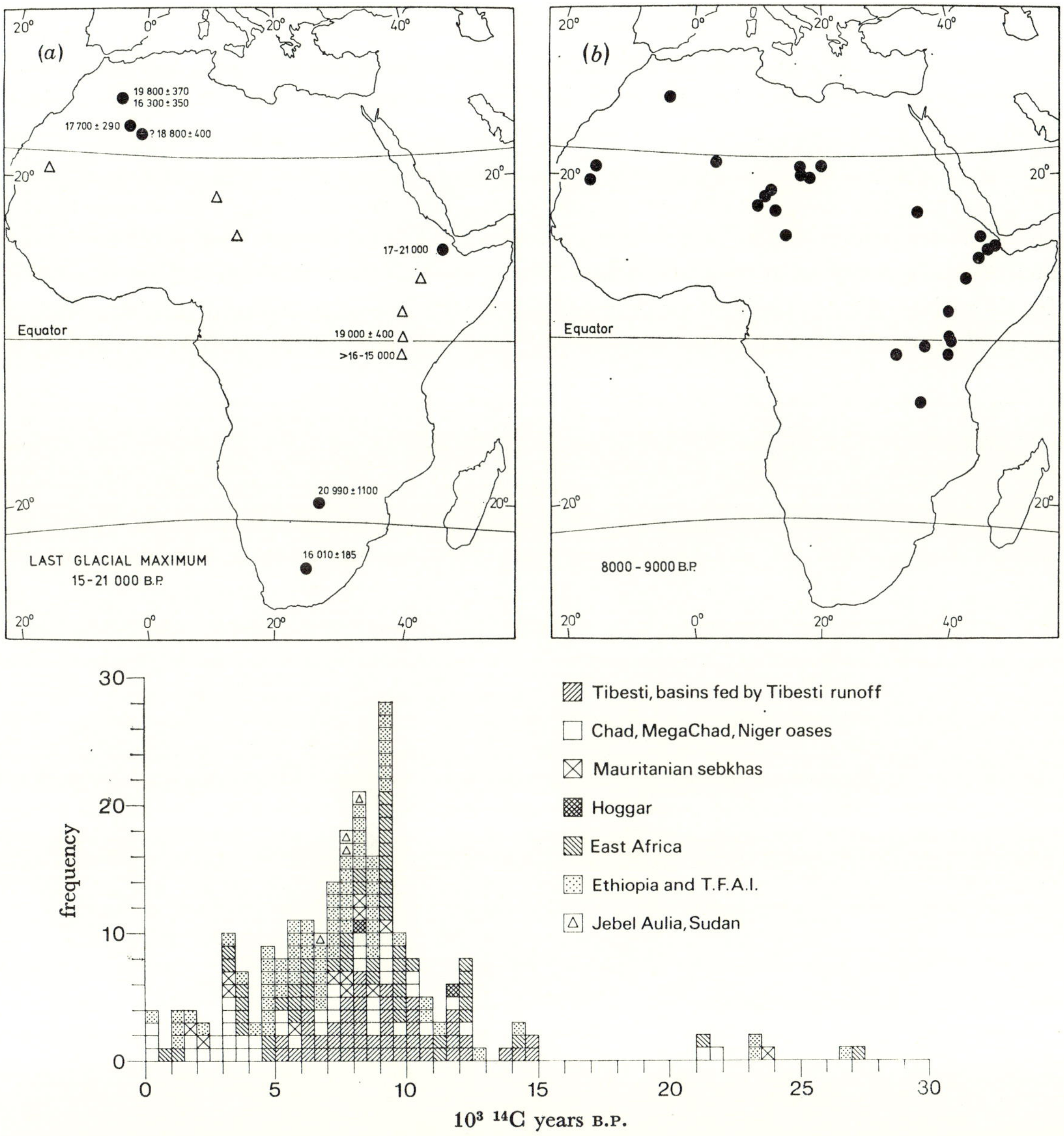

FIGURE 3. (*a*) Africa about 18000 B.P. Circles indicate dated lacustrine sediments, triangles that the lakes were dry at this time. (*b*) Africa 8000–9000 B.P. with dated lacustrine sediments, circles, indicating the presence of lakes in numerous tropical basins at this time. (*c*) Dated lacustrine sediments indicating that lakes stood at or above present levels (from Street & Grove 1976).

most of the surface within 40° of the Poles; loess formations were accumulating in middle latitude, semi-arid regions of the present day, and although there were extensive lakes in southwest United States of America and other mid-latitude regions now semi-arid, they were the result of lower temperatures and evaporation rather than increased precipitation (Flint 1971). In general, the inter-tropical regions were drier than at present (Williams 1975). Active dunes occupied much of the Sudan zone of Africa (Grove 1958), the semi-arid regions of north-west India (Goudie 1973), North-Central Australia (Bowler *et al.* 1975) and savannalands in the basin of the upper Orinoco and Sao Francisco rivers of South America (Tricart 1975). Arid conditions thus extended into regions now semi-arid.

After about 15000 years ago, global climatic conditions ameliorated, with precipitation increasing and temperatures rising, though not steadily. By 9500 years ago, basins in tropical Africa held very extensive lakes (Street & Grove 1976; figure 3). Temperatures, except in the vicinity of the Laurentide ice sheet, were similar to those of the present day, and in tropical Africa it would seem that annual rainfall was about 150 % of present day means. Semi-arid conditions in Africa and possibly over much of the globe were limited to certain west coastal strips in low latitudes and to regions that are now arid. The present semi-arid regions were much better watered than they are today. Lakes now confined to closed basins overflowed; Megachad into the Benue in West Africa (Pias 1970); the Galla lakes of Ethiopia into the Awash; Lakes Stefanie and Turkana to the White Nile (Grove, Street & Goudie 1975).

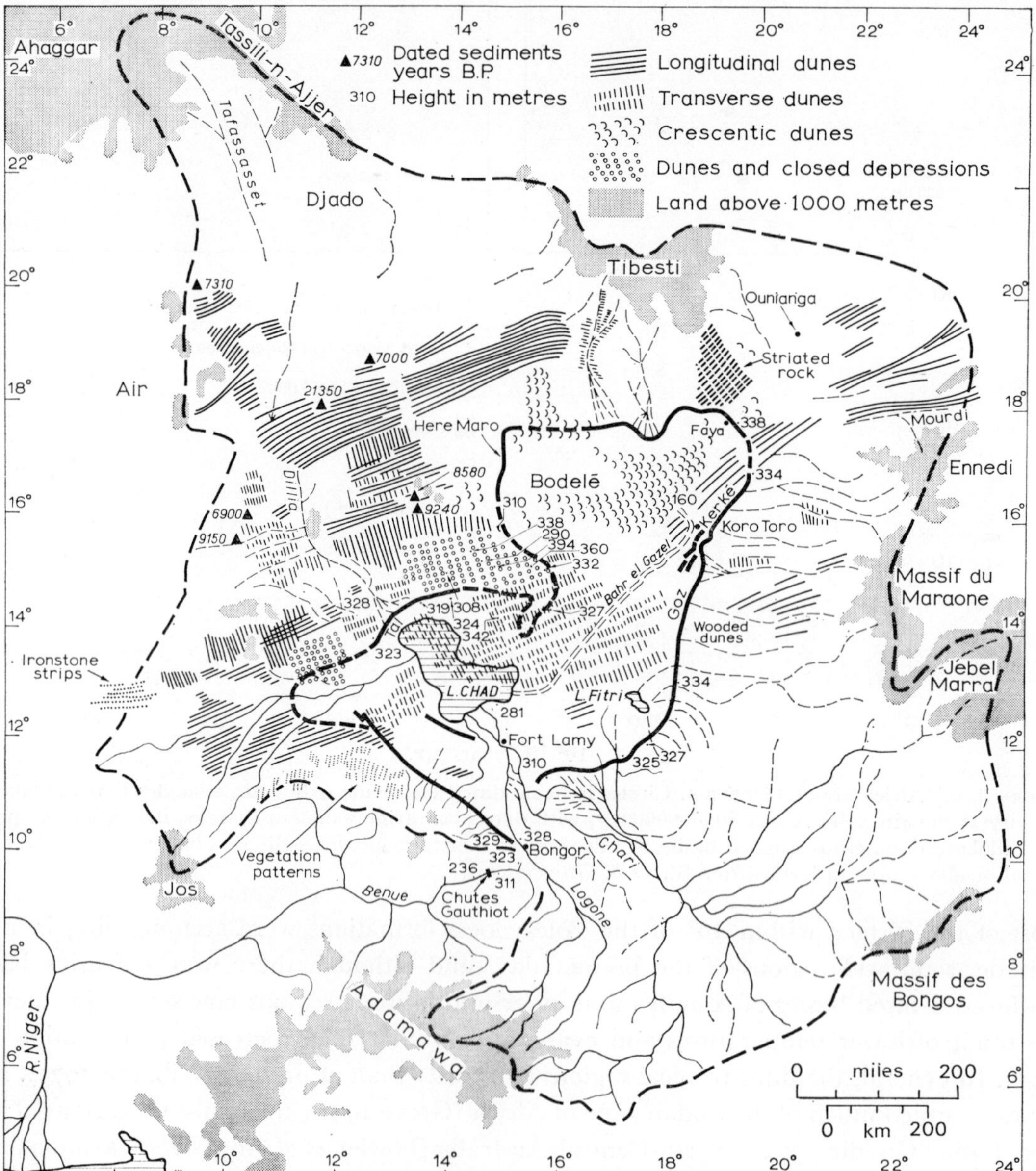

FIGURE 4 (*a*). For description see opposite.

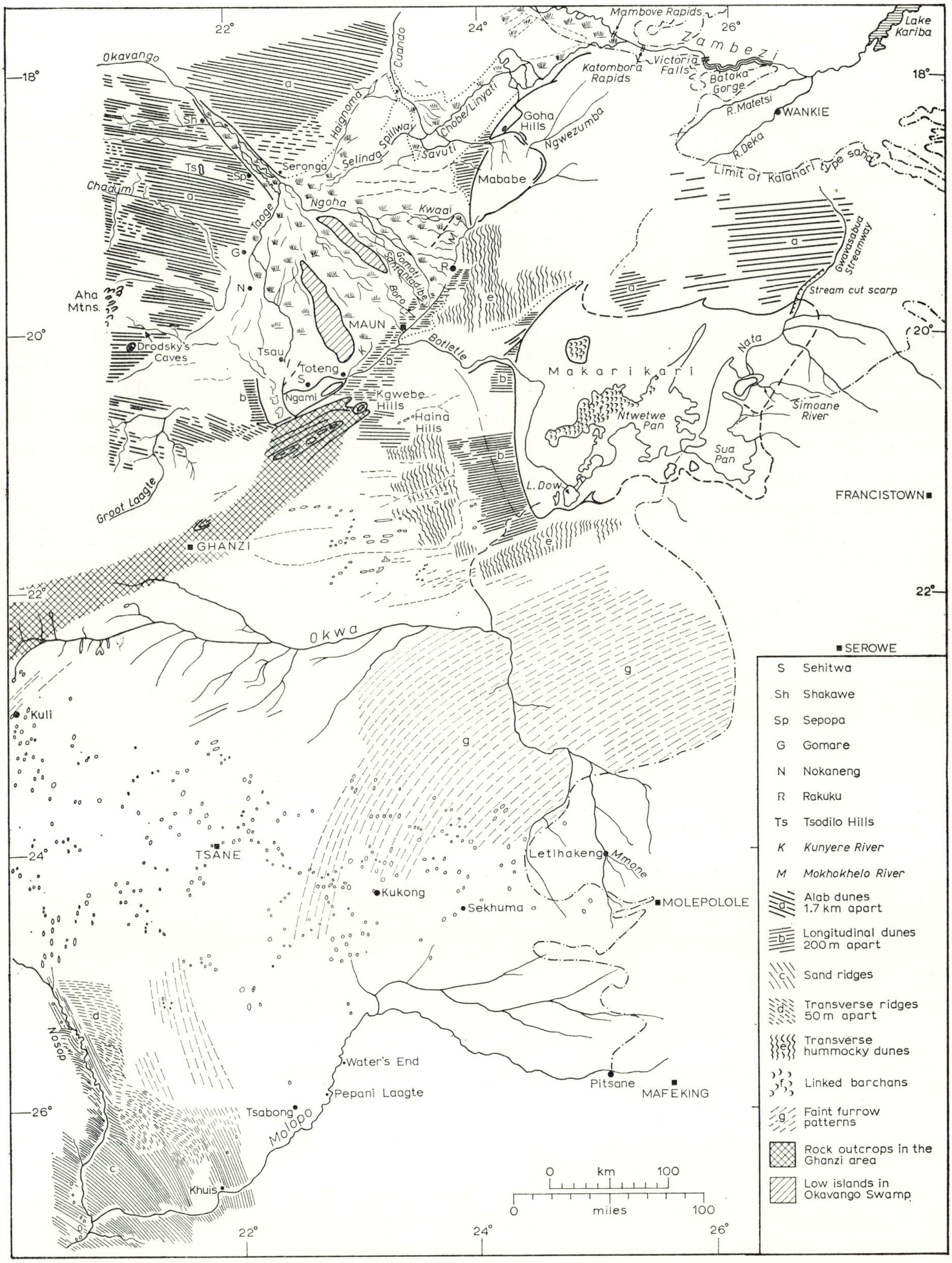

FIGURE 4 (b). (a) Late Quaternary dunes and lakes in the Chad basin (from Grove & Warren 1968). (b) Late Quaternary dunes and lakes in the Kalahari (from Grove 1969).

About 7000 years ago climates began to deteriorate, fluctuating from time to time, but after 5000 B.P. becoming somewhat cooler and, in tropical Africa at least, markedly drier. Since 4500 B.P. there have been fluctuations of climate but on a smaller scale than those of the preceding 15000 years.

The climatic changes of the last 20000 years, that have been briefly summarized above, are only the latest in a Quaternary history extending more than a million years into the past. From our point of view these latest changes are important because of the deep impression they have made on the semi-arid lands as we find them today. To an observer on the ground the traces left by them may not be immediately apparent, but from the air and from satellites they are readily recognizable. The land surfaces of semi-arid Australia, the Saharan borderlands, the Kalahari, northern Argentina, the southwest U.S.A. and central Asia preserve features derived from both the wetter and the drier episodes of the Late Quaternary (figures 4 *a*, *b* and 5, plate 1). Soils are closely related to the sedimentary landforms with soil properties varying greatly over short distances according to the sediments underlying them and the morphology of landforms resulting from past wind and water action. Various organizations concerned with mapping the resources and assessing the potential of extensive tracts of semi-arid country have used the morphology and patterns appearing on air photographs as aids to ground surveying. One might mention the publications of the C.S.I.R.O. in Australia, the French O.R.S.T.O.M. on the south side of the Sahara, the Land Resources Division of the United Kingdom Directorate of Overseas Surveys in Botswana and Nigeria, Huntings and other commercial organizations in Sudan and Iraq among other countries.

In tropical semi-arid regions, soils derived from dunes that were active in the last glacial period, though leached and even gullied in the succeeding period of high biological activity of Early Holocene times, are productive under cultivation; they are particularly suitable for crops like groundnuts and bullrush millet. In the basins of Chad and on the plains east of the White Nile in Sudan, heavy montmorillonitic clay soils, derived from riverine, lagoonal and lacustrine clays, are of increasing agricultural importance. They are difficult to cultivate with traditional tools, but deep cracking in the dry season reduces weed growth and certain varieties of sorghum have long been grown, being sown towards the end of the rains, drawing on water stored in the soil and harvested in the middle of the dry season. The clay soils of the Gezira are derived from sediments of the Blue Nile laid down at this time of high discharge in the Early Holocene. Both the dune and the lacustrine clay soils are more fertile than older soils in the same general regions, which have been strongly leached over a long period of time and commonly include lateritic ironstone.

On the poleward sides of the tropical deserts, in the Kalahari, on the northern margins of the Sahara, in the southwest U.S.A., and in parts of Australia, lime has accumulated in soil profiles and may exist as hard caliche and calcrete. Its development is also believed to be related to the climatic fluctuations of the Late Quaternary. In higher latitudes, towards the margins of the glaciated regions, the chernozems and other soils of the grasslands and steppes are commonly derived from loess and other wind-deposited or alluvial materials of Late Quaternary times. Among them are some of the world's most productive wheatlands.

The availability and reliability of subsurface water in the deeper aquifers of semi-arid lands, especially in low latitudes, is related to past climatic conditions. They were charged in the more humid periods of the past and may no longer be replenished. Use of such water is a form of mining, the consumption of a non-renewable resource.

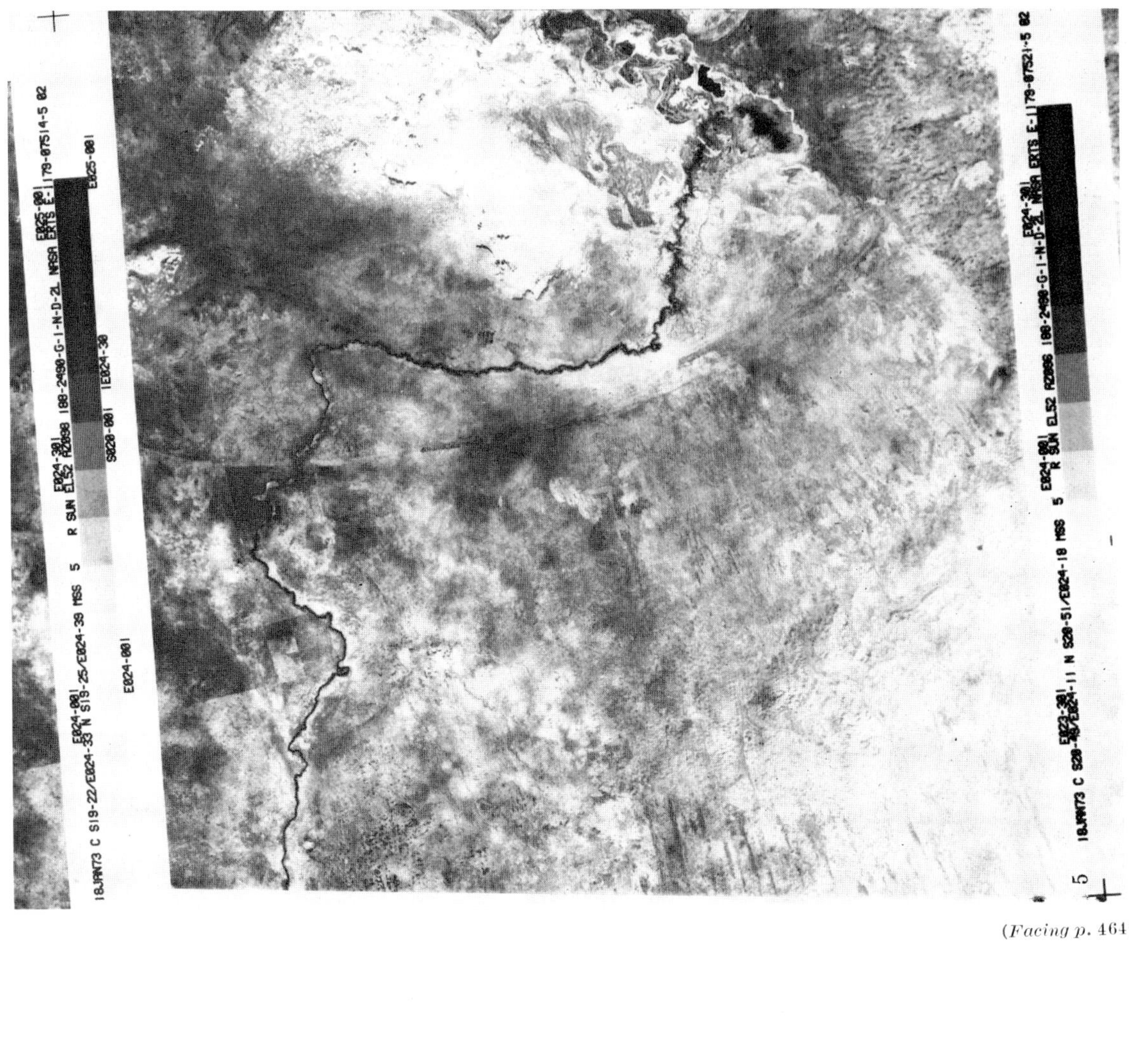

5

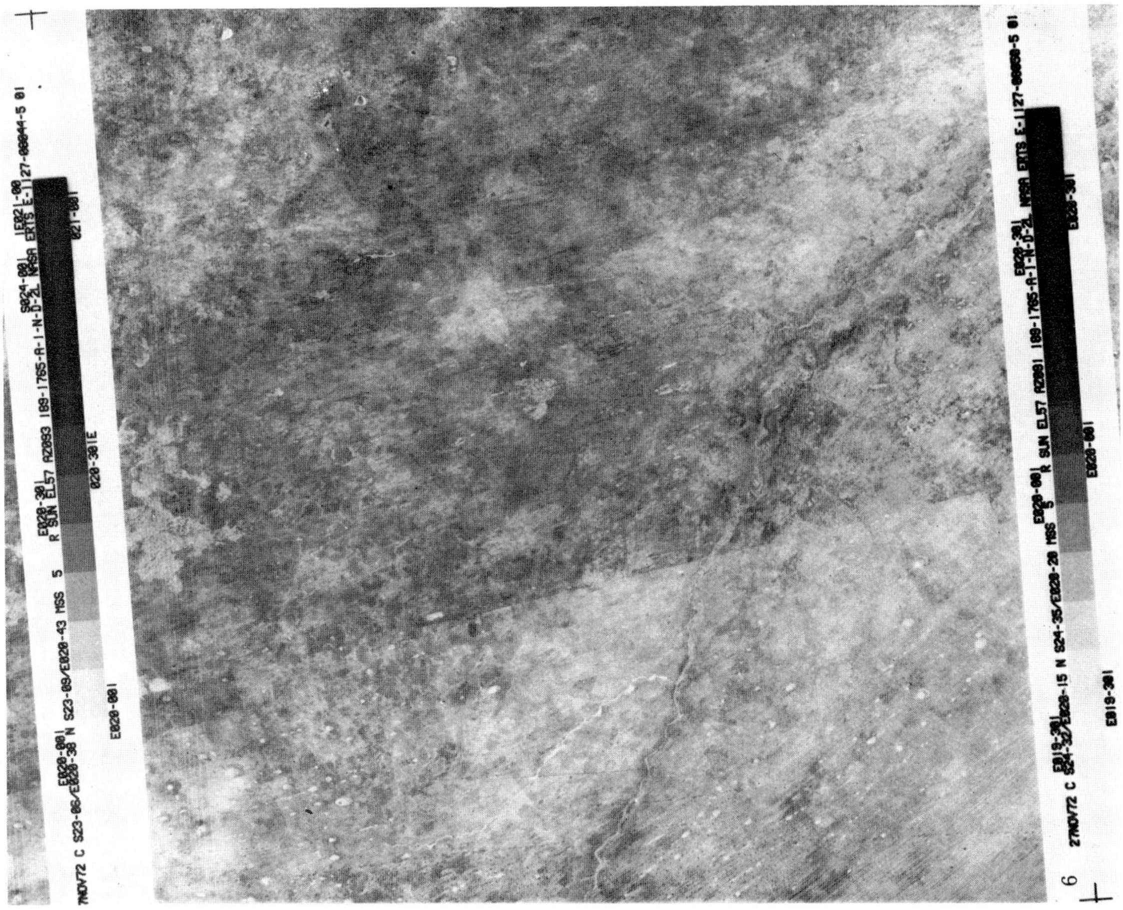

6

FIGURE 5. ERTS. The western margin of Pleistocene lake in the Makarikari depression, Botswana. 1179–07521.

FIGURE 6. ERTS. Vegetated dunes and pans on either side of the frontier between Botswana and Namibia. 1127–08050.

(*Facing p.* 464)

3. Modification of semi-arid ecosystems by man

The fauna and flora of semi-arid lands are adapted to water stress in various ways. In general, biological productivity is low as compared with temperate and low latitude forested regions on account of seasonal lack of water, and the frequency of disturbances such as floods and droughts. Much of the biomass is beneath the surface in the form of roots and bulbs. In low latitudes the number of species present is low, especially at heights above 1500 m. Generally, plant systems and communities are simple and robust, but life forms vary markedly over short distances according to the availability of water on sites differing in slope, soil depth and drainage.

The proportion of the surface of the ground covered by transpiring vegetation surface diminishes as aridity increases, but the amount of cover may vary greatly over short distances even on gently sloping plains with rather undifferentiated soils. Many such areas with mean annual rainfall totals of 250–750 mm are characterized by rhythmic banding of the vegetation; the *brousse tigrée* of southern Mali and Niger, vegetation arcs in Sudan, Somalia and Jordan, the groves and lanes north of Alice Springs in central Australia. These features have received a good deal of attention but the patterning is still not well understood. Its effect is to concentrate the available water in a small proportion of the total area, thereby allowing higher forms of life to persist than would otherwise be the case.

The density of the vegetation cover has everywhere in semi-arid lands been modified by man, in most places throughout the course of the Holocene. In recent decades the disturbance resulting from grazing, and general use by growing populations, has been greater than ever. Its severity has varied from one country to another and in certain regions the contrasts are so marked that international frontiers can be recognized on satellite imagery from the tonal differences (figure 6, plate 1).

It is possible that the increased albedo has strengthened subsidence of air over those regions where the vegetation has suffered most, thereby intensifying the aridity, but this is so far unproven (Charney 1975). There is no doubt that exclusion of people and their stock would allow regeneration of the vegetation, and it is possible to sow and plant indigenous and exotic species of grasses and trees, but such measures are difficult politically and generally too costly to undertake.

In recent years efforts to conserve and rehabilitate semi-arid land have been directed to those areas that have been irrigated in the past and where salt accumulation has damaged the soil. Now attention is turning to the vast areas threatened by the extension of desert conditions in the Sahel and Sudan zones and in other similar parts of the world. The problems here are immense, and solutions may involve attracting people from such areas into towns and into regions more able to sustain population pressure. Perhaps it would be more rewarding to concentrate attention on safe-guarding areas of special ecological significance such as the inland deltas. In Africa the Sudd, Okavango, inland Niger and Logone-Shari deltas are rich floristically and faunistically but liable to drastic modification as a result of development. It might also be argued that it is in such areas subject to flooding that responses to reliable climatic predictions might most usefully be developed.

The fauna of semi-arid lands was once remarkably rich in large species of herbivores and the creatures that preyed on them. The large grazing animals are walking energy stores, able to move long distances and make good use of primary production varying greatly with the season and from place to place. A century ago, the kinds of game numbers that still exist in Serengeti were encountered over much more extensive tracts of semi-arid Africa. Cloudesley-Thompson (1967) has collected together many accounts by travellers in earlier centuries, some of them responsible for introducing the precise firearms that have reduced numbers so drastically.

While the numbers of the larger animals have been drastically reduced, conditions for the survival of certain species of the smaller ones, especially the smaller social creatures, like quelea and locusts, have been improved in some respects, with irrigated land providing oases where they can rest and feed. Control of locusts appears to have been increasingly successful and in recent years the threat of quelea and rats has been more serious to countries in semi-arid Africa.

4. The history of development of semi-arid lands

Pastoralism probably evolved in semi-arid lands over the millennia as symbiotic relationships developed between people and certain species of animals. Now most semi-arid land is devoted to pastoralism, although its importance in terms of the numbers of people involved in pastoral production and its value in these regions is commonly exceeded by agriculture or other activities. Generally the seasonal variations in the availability of water and grass result in nomadism or, more commonly, regular seasonal migrations comparable to those of antelope and other wild animals. Traditional movements have been modified by international frontiers and by interference from governments mainly responsive to pressures from city-dwellers (Monod 1975; Leshnik & Sontheimer 1975).

The harvesting of grasses and the sowing of cereals originated in semi-arid lands, probably in the lands around the eastern Mediterranean in the Late Quaternary. Cereals like sorghum were probably domesticated on the south side of the Sahara about the end of the last humid period. The desert and its semi-arid margins continued to provide the setting for later civilizations, Persian and Harrapan, Greek and Roman in North Africa, the Islamic world extending from Morocco to India, the Empires of Ghana, Mali and Songhai in the western Sudan (Stamp 1961).

The environment was always defective and its costs therefore high because of the shortage of water, except near major rivers rising in wetter areas. Wood was scarce for constructional purposes, fuel and charcoal, and became scarcer as time went on. Forests in neighbouring areas with higher relief and rainfall were exploited with resultant damage to the vegetation, soils and rivers of the mountains and accelerated sedimentation on valley floors in the semi-arid lands. Desert extended into the lands along the south and eastern sides of the Mediterranean, and the winter rainfall zone was degraded. Shortage of wood in the region may help to explain why it fell so far behind Europe in the 17th and 18th centuries. Now, concrete and plastics, petroleum and natural gas help to offset the shortage of wood.

Much uncertainty exists as to the timing of the deterioration of the plant cover around the Mediterranean and elsewhere. But it is clear that the danger of desert conditions spreading into semi-arid lands has never been greater than in this century, with numbers of people and livestock increasing at unprecedented rates, and the individual's demands on environmental resources increasing still faster. Everywhere the vegetation cover of semi-arid lands has been degraded by burning, grazing, felling and cultivation, thereby setting in train soil deterioration.

Erosion losses are naturally high under semi-arid conditions because of the relatively poor protection afforded by the vegetation and the occasional occurrence of heavy storms. With biome changes, erosion by wind and water accelerate. Sediment yield per unit area, already high, increases still further, overloading rivers with coarse sediments, clogging channels, and transforming meandering rivers into braided waterways while headwater streams are entrenched. Commonly, road construction works and culverts, by concentrating runoff from wide areas, can initiate channel erosion and the formation of arroyos.

Reservoirs in semi-arid lands require extensive catchments; with high sediment yields per unit area they are liable to fill rapidly. Once they are rendered useless, alternative dam sites are difficult to find and costs of water storage mount as time goes on.

The solute load of rivers in semi-arid lands depends on conditions in the source regions. The Niger and Senegal rivers rising on the lateritized high plateaus of West Africa are so dilute that when their waters are used for irrigation, soils are liable to become increasingly leached and acid. The waters of the Blue Nile, rising on lava highlands in Ethopia, are much richer in nutrients and very suitable for irrigation (Grove 1972). Some other rivers rising in the mineral springs of tectonically active regions, or on rocks containing evaporites, have excessively high solute loads. Entering semi-arid country from more humid highlands, such rivers acquire increasingly high sodium/calcium ratios and, if such rivers are used for irrigation, risks of soil salinization are high (Gibbs 1970). Drainage water from irrigated land enhances the solute load of rivers already carrying much material in solution. Industrial and urban wastes may further overload the hydrological system.

Underground sources of water, some of them artesian or sub-artesian, have been heavily exploited for watering stock, irrigating crops and for industrial and domestic purposes. Almost invariably the large volumes required for irrigation lower water tables, necessitating the deepening of wells and boreholes and the installation of more powerful and expensive pumps. The needs of traditional pastoralists and villagers are usually met without too much difficulty, for the essential needs of a single person or livestock unit are only about 25 l per day and a single well yielding 5000 l an hour can supply the needs of a village or the cattle likely to be within daily walking distance of it. But such a well could provide enough water to irrigate only a hectare or two in the dry season. In western Texas, according to Dregne (1970), '63000 irrigation wells pump from 4900 to 9800×10^6 m^3 of water from the Ogallala formation from depths of 30 to 90 m, to irrigate 2.1 M hectares of land. Since 1949 the average pumping lift in Hale County, Texas, has increased from 12 to 30 metres.' People living in urban areas use several times as much water as villagers and the demands of industrial plants are enormous.

In spite of the obstacles to development in semi-arid lands, the costs resulting from water scarcity, and the hazards of drought, they have always been attractive to people. Stores of energy and nutrients have accumulated in the grassland soils; the vegetation cover is readily cleared for agriculture; abundant sunshine makes for pleasant living conditions. Particularly well-favoured are the coastal semi-arid lands of middle latitudes. With winter rain and then sunshine, a wide range of crops can be grown, both annuals like cereals and pulses, and perennial citrus, vines, olives, figs and dates. With water from underground sources or rivers from more humid neighbouring regions, such crops yield more heavily than anywhere else in the world.

Modern technology has done much to improve living conditions in semi-arid lands. Air conditioners temper the heat of summer, and refrigeration is a boon. The most attractive areas have been coastal semi-arid lands backed by mountains. Piedmont areas in the continental interiors are also booming, cities in such regions having very high growth rates. Manufacturing industry, using water from the mountains and underground, is expanding. People like to live in such environments so that labour is no problem and tourism is important. Additional economic resources may be associated with the climate, notably fisheries harvesting the product of upwelling ocean waters offshore. Other resources of a more incidental kind, notably phosphate and mineral oil, have stimulated economic growth in certain instances.

Human use has involved deterioration in the habitats of semi-arid lands especially in the last few decades. Stores of underground water and soils built up over long periods of time have been depleted in a few decades in the course of this century. The costs of restoring the situation, by pumping in water, by desalinization or by afforestation are inevitably high. Productivity can be restored and the semi-arid lands be made to yield more crops and livestock and support more people in a pleasanter environment, but only with heavy expenditure in skill and equipment.

5. Development of semi-arid lands

The future of semi-arid lands will vary according to the wealth of the countries in which they are situated. Some of the dry lands are remote, distant from the coast, from ocean transport and contacts with the rest of the world. Some are of only marginal value to the states of which they form a part, others are of vital importance to the nation. The states themselves vary in size and population, wealth and stage of modernization. All these influence the prospects for development and the availability of funds to finance it. The long-term effectiveness of such development depends heavily on the management skills of governments and other organizations involved in the process.

A useful indicator of the funds available locally for investment in semi-arid lands is the gross domestic product (g.d.p.) and g.d.p. per head of the population. Out of some 50 countries with sizeable areas of semi-arid land, some like Mali, Niger and Upper Volta, Sudan, Ethiopia and Somalia are among the poorest in the world, with g.d.ps per capita of less than $200 per year. A small number are wealthy, with over $2000. These last fall into two groups, industrial countries and those with infant industries but rich mineral resources, notably mineral oil and natural gas. There are other countries with extensive areas of semi-arid land, including China and India, which do not fall neatly into either group.

(a) *Development in industrial countries*

In the wealthy, industrialized countries, development of semi-arid land has usually involved, first, extensive pastoralism and ranching, then, more locally, mining, dry farming and irrigation. In Northern Australia, tourism now exceeds livestock production in value and both are far exceeded in importance by mining.

Oil prospecting and the development of oilfields and pipelines has led to the discovery and exploitation of underground water resources, notably in the borderlands of the northern Sahara. In South Africa, water pumped from the deep goldmines of the Witwatersrand adds significantly to local supplies. But industry and mining usually compete with agriculture for scarce supplies of water in semi-arid lands. For the production of a tonne of maize, 5000–10000 m^3 of water may be needed, enough to allow the production of, say, 20 tonne of steel. Where industry and agriculture compete for water, as in the Vaal river basin, it is the needs of industry that usually turn out to be the greater, in financial terms.

In Australia, South Africa, the U.S.A. and U.S.S.R., semi-arid lands are in close proximity to industrial heartlands. Especially in the U.S.A., urban expansion has been rapid, with towns attracting manufacturing industry, and also large numbers of retired people. Costs of development are high, and it has been argued that such urban growth is effectively a drain on the national economy rather than providing a net contribution to it. It could be replied that using the semi-arid lands in this way adds to the enjoyment of the nation even if it does not add materially to its economic benefits. Nevertheless, it is possible that only wealthy countries can afford to direct investment to the growth of industrial centres in semi-arid lands, very dependent on inputs of resources and finances from outside, and possibly contributing less than they receive to national economies.

TABLE 1. INDUSTRIAL COUNTRIES

	(*a*) area 10^6 km²	(*b*) pop. 10^6	(*c*) *d* pop. (% per year)	(*d*) g.d.p. 10^6 \$	(*e*) g.d.p./head \$/year	(*f*) agric. g.d.p./head \$/year	(*g*) oil 10^6 tonnes per year	(*h*) dry area (%)
Africa								
South Africa	1.2	23.0	2.7	19928	843	–	–	55
N. America								
U.S.A.	9.3	209.8	1.1	1159318	5551	2285	467	35
Canada	10.0	21.9	1.6	104979	4805	1311	73	4
Mexico	2.0	52.6	3.5	41096	781	153	21	52
S. America								
Argentina	2.7	24.0	1.5	29706	1261	646	22	54
Asia								
Israel	0.02	3.1	2.9	7077	2298	785	6	75
U.S.S.R.	16.8	61.7	1.7	–	–	–	–	–
Australasia								
Australia	7.7	13.0	1.9	41098	3228	2092	16	82

The figures in all the tables are for 1972 unless otherwise indicated.
Sources: (*a–e, g*) *United Nations Statistical Yearbook*, 1973. (*f, h*) F.A.O. Committee on Agriculture: *Improving Productivity in Low Rainfall Areas* (COAG/74/4, Rev. 1, April 1974, pp. 9, 25).

The development of agriculture in the semi-arid lands of industrial countries presents hazards of two kinds for the future. Firstly there is the risk that water tables will be lowered to such an extent that supplies for irrigation will diminish. Such risks are predictable and therefore of less significance, certainly to the rest of the world, than the risks of failure of rain-fed crops as a result of drought. The importance of shortfalls in cereal production, notably in the U.S.S.R., are generally appreciated. They played a part in the increase in commodity prices in 1972 and deficits in 1975 are attracting attention currently. So long as North American harvests are satisfactory, global demands can be met, but if drought were to reduce production in America at the same time as in the U.S.S.R., and especially if drought recurred in successive years, the effects on world food stocks and the financial system would be severe. The rôle of these semi-arid lands in international economic affairs is clearly a very important one.

(*b*) *Development in oil-rich countries*

Of particular importance today are the semi-arid regions of those countries which have only recently begun to industrialize and where the extraction of minerals, especially petroleum, is attracting enormous sums of money from the oil-consuming countries; western Europe, the U.S.A. and Japan, with their great industries and wealthy populations. Part of the money is being invested in the oil-consuming countries, a part in purchasing consumer goods and armaments, a third part is being spent on capital goods with the aim of installing utilities of various kinds and manufacturing plants. Fourthly, a part is being directed towards the poorer countries, many of which are Islamic and semi-arid.

About 60 % of the world's oil reserves and a half of present production are concentrated around the Persian/Arabian Gulf. Some of the states involved are small and entirely arid and do not therefore enter into our considerations directly. The larger ones include great tracts of semi-arid land. Much of Iran's $20 000 million per year income is being directed into industry

Table 2. Oil rich countries

	(*a*) area 10⁶ km²	(*b*) pop. 10⁶	(*c*) *d* pop. (% per year)	(*d*) g.d.p. 10⁶ $	(*e*) g.d.p./head $/year	(*f*) agric. g.d.p./head $/year	(*g*) oil 10⁶ t/year	(*h*) dry area (%)
Africa								
Algeria	2.4	15.3	3.5	4156[69]	299[69]	52	51	96
Libya	1.8	2.1	3.7	4860	2337	79	106	100
Nigeria	0.9	58.0	2.5	5310[69]	99[69]	61	91	28
S. America								
Venezuela	0.9	11.0	3.4	14159	1291	224	168	9
Asia								
Iran	1.6	30.6	3.0	17001	556	130	248	85
Iraq	0.4	10.1	3.3	3141[69]	343[69]	106	71	97
Oman	0.2	0.7	3.0	–	–	–	14	100
Saudi Arabia	2.1	8.1	2.8	–	–	48	286	100

and agriculture in semi-arid land. With a population of 30 million growing at 3 % annually, Iran can absorb such sums. Both Algeria and Iraq are in a similar position to Iran. Oil in Algeria brings in about $3700 million per year; it has a population of over 15 million growing at 3.5 % annually and is investing mainly in its Mediterranean fringe, much of which is semi-arid. Iraq's population of more than 10 million is growing at 3.3 % annually and its oil income of $6700 million can be invested locally. Nigeria's $8000 million are being spent widely through the country, the greater part of which is outside the arid zone, but considerable sums are being invested in irrigation schemes in the semi-arid northern districts, where pressure of population on land resources has for long been high.

In all these oil-producing countries, expenditure first by oil companies and latterly by national and regional governments on roads, urban facilities and the civil service is attracting the traditional pastoralists and the sedentary cultivators from the countryside. Lack of rain in individual years pushes more people into the towns, not only from oil-producing countries but also from neighbouring territories. Everywhere, g.d.p. per head in agriculture is well below the mean national figure and the discrepancies have been magnified greatly over the last three or four years. The exodus from the countryside to the towns will inevitably accelerate (Clarke 1973).

So far, investment in agriculture has not been outstandingly effective. Rising food prices have been associated with increased imports of foodstuffs from overseas. Eventually, one may suppose, incomes of cultivators and their ability to farm more profitably will improve, but they have not always been helped by governments subsidizing imported foods to keep down the costs of living for the townspeople.

The situation is different in the oil-producing states with small populations. Kuwait falls into this class as do a number of the other Gulf States. Some funds from these countries have moved in the direction of the Sudan Republic for agricultural development. Both Saudi Arabia and Libya have very large incomes from oil in relation to their populations and the land at their disposal that is not arid. Saudi Arabia's 8 million people, increasing at 2.8 % annually, have $30 000 million annually at their government's disposal; Libya, with 2 million people, has an oil income that varies but is of the order of $6000 million per year. In both countries nomadic pastoralism has declined and intensive stock-rearing using imported feeding stuffs is starting. In agriculture, much of the investment in land and irrigation is by townspeople and is in the semi-arid areas adjacent to urban areas. The spread of irrigation in the semi-arid northwest coastal regions of Tripolitania has resulted in a rapid fall in the level of groundwater.

In the short term it might be argued that the chief problems to be faced are social and political. In Saudi Arabia, much reliance is placed on Yemeni labourers to carry out the more menial tasks, Libya depends on Tunisians, and Kuwait on Lebanese and other nationalities. Where oil money can provide for all, and everyone is a lily of the field, will society survive unscathed? Fundamentally, the industrial west is concerned with the development of semi-arid lands because it is to them that a large part of the payments they make for oil are being directed, and this involves the product of their factories, farms and labour forces. Ineffective investment represents a waste of the world's resources and of human effort.

(*c*) *Development in the poor countries*

The problems of resource development in semi-arid lands are most severe where the countries concerned are entirely in the arid zone and where their mineral resources are on a modest scale or non-existent. The Sahelian countries typify this situation. Senegal, Mali, Upper Volta and Niger with a total population of about 20 million have per capita g.d.p.s of about $200 and in agriculture of about half that amount. Traditional systems of pastoralism persist, and the cycle of poverty, lack of investment and no improvement in living standards is hard to break. In the years from 1968 to 1974, gross production rose slowly but production per head of population fell. Wealth in the form of cattle was destroyed and the process of recovery will take several years without any guarantee that similar setbacks will not recur with increasing frequency as populations continue to increase. All four countries, except Senegal, are land-locked. Upper Volta has no navigable rivers; the other countries include stretches of the Niger and Senegal with their flood plains which provide opportunities for large scale irrigation schemes and power development. International river commissions have been in existence for some years and various schemes for developing the country bordering the southern parts of lake Chad have been prepared. It seems likely that Nigeria, with funds from oil and the need for land, will go ahead with its own agricultural schemes in that part of the Chad basin lying in Borno State without paying too much regard to the other countries concerned.

In southern Africa, both Botswana and Namibia are entirely in the arid zone. Both have larger mineral resources than the states of the western Sahel and their economies are associated with South African industry. Both have small populations, and prospects for the future are better than in the countries of the Sahel.

Sudan and Ethiopia include much country outside the arid zone. Vast areas of clay plains in the Sudan, east of the White Nile, have been brought under the plough in recent decades and fears have been expressed as to whether fertility can be maintained. Attention is now being directed to the rainlands of the Qoz, the fixed dune country west of the Nile and north of the Bahr-el-Arab where there are opportunities, in the savanna lands of southern Darfur, to work out schemes of sustained yield agriculture before the population reaches critical levels.

In Ethiopia, the most extensive areas of semi-arid land lie on the plateaus of the southeast, with soils derived in part from volcanic rocks. Wide stretches have been brought under cultivation in recent decades and opened up by new roads. The opportunities for increasing the productivity of these lands would seem to be good.

TABLE 3. POORER COUNTRIES. MAINLY ARID AND SEMI-ARID

	(*a*) area 10^6 km²	(*b*) pop. 10^6	(*c*) pop. (% per year)	(*d*) g.d.p. 10^6 \$	(*e*) g.d.p./head \$/year	(*f*) agric. g.d.p./head \$/year	(*g*) oil 10^6 t/year	(*h*) dry area (%)
Asia								
Afghanistan	0.6	17.9	2.3	1496^{70}	88^{70}	44	0	81
Jordan	0.1	2.5	3.4	671	272	108	0	98
Pakistan	0.8	56.1	2.4	11186	204	60	0	90
Syria	0.2	6.7	3.4	2277	341	118	6	83
Yemen (Arab Rep.)	0.2	6.1	–	–	–	35	0	92
Yemen (Dem. Rep.)	–	–	–	–	–	–	0	100
Africa								
Ethiopia	1.2	25.9	1.9	–	–	38	0	74
Morocco	0.5	15.8	–	3872	245	95	0	85
Somalia	0.6	2.9	2.3	231^{69}	85^{69}	–	0	100
Sudan	2.5	16.5	2.7	1832	117^{70}	60	0	91
Tunisia	0.2	5.4	3.0	1609	304^{71}	76	4	92
Africa								
Botswana	0.6	0.6	2.8	140^{71}	209^{71}	–	0	91
Chad	1.3	3.2	2.0	270	74^{70}	33	0	92
Kenya	0.6	12.1	–	1992	165	47	0	75
Mali	1.2	5.3	2.0	262^{69}	54^{69}	39	0	95
Mauritania	1.0	1.2	2.2	182^{69}	160^{69}	50	0	100
Namibia	0.8	0.6	2.0	–	–	–	0	90
Niger	1.3	4.2	2.6	378^{69}	97^{69}	56	0	100
Senegal	0.2	4.1	2.4	1029	250	79	0	87
Upper Volta	0.3	5.6	2.1	–	–	26	0	94

TABLE 4. POORER COUNTRIES. PARTLY ARID AND SEMI-ARID

	(*a*) area 10^6 km²	(*b*) pop. 10^6	(*c*) *d* pop. (% per year)	(*d*) g.d.p. 10^6 \$	(*e*) g.d.p./head \$/year	(*f*) agric. g.d.p./head \$/year	(*g*) oil 10^6 t/year	(*h*) dry area (%)
Africa								
Angola	1.2	5.8	–	–	–	49	7	23
Cameroon	0.5	5.0	–	1091^{70}	187^{70}	84	0	9
Madagascar	0.6	6.2	–	973	141^{71}	–	0	15
Tanzania	0.9	14.0	2.6	1529	112	38	0	20
S. America								
Brazil	8.5	98.9	2.9	50717	513	98	8	5
Bolivia	1.1	5.2	2.6	1106	219	56	2	22
Chile	0.8	8.8	–	5684	574	166	2	47
Paraguay	0.4	2.6	3.4	769	298	141	0	8
Peru	1.3	14.4	3.1	5473	378	99	3	17
Asia								
India	3.3	563.5	2.2	49087^{69}	96^{69}	–	7	–
Turkey	0.8	37.0	2.5	16121	436	136	3	41

Many other countries so far not mentioned by name have their own particular problems and possibilities for development. Some fall neatly into one or other of the categories that have been distinguished, others have the characteristics of more than one of the conditions affecting the development of their resources.

The conclusion of a working group that recently reported on the obstacles to development of arid and semi-arid lands was that 'the overarching need is for the refinement of integrated approaches to the planning and development programmes' and that 'most efforts have suffered from concentration on narrow sectors to the detriment of integrated approaches that take in the whole complex of the resources and human needs of a region and that look to practical use of the findings' (U.N. Economic and Social Council).

References (Grove)

Aspin, D. 1976 Rainfall in the Seychelles 1941–70. *Weather* **31**, 47–56.

Bowler, J. M., Hope, G. S., Jennings, J. N., Singh, G. & Walker, D. 1975 Late Quaternary climates of Australia and New Guinea. Proceedings of WMO/IAMAP Symposium on long-term climatic fluctuations, Norwich 1975, WMO No. 421.

Caviedes, C. N. 1975 El Niño 1972: its climatic, ecological, human and economic implications. *Geogr. Rev.* **65**, 493–509.

Charney, J. G. 1975 Dynamics of deserts and drought in the Sahel. *Q. Jl R. met. Soc.* **101**, 193–202.

Chu Ko Chen 1973 A preliminary study on the climatic fluctuations during the last 5,000 years in China. *Scient. Sin.* **16**, 226–256.

Clarke, J. I. 1973 Urban population growth in the Middle East and North Africa. In *Maghreb and Sahara* (ed. Xe de Planhol). Société de Géographie. Paris:

Cloudesley-Thompson, J. L. 1967 *Animal twilight.* London: Foulis.

Dov Nir 1974 *The semi-arid world.* London: Longman.

Dregne, H. E. (ed.) 1970 *Arid lands in transition.* Washington D.C.: American Association for the Advancement of Science.

Flint, R. F. 1971 *Glacial and Quaternary geology.* New York: Wiley.

Gibbs, R. J. 1970 Mechanisms controlling world water chemistry. *Science, N.Y.* **170**, 1088–1090.

Goudie, A. S., Allchin, B. & Hegde, K. T. M. 1973 The former extensions of the Great Indian Sand Desert. *Geogr. J.* **139**. 243–257.

Grove, A. T. 1958 The ancient erg of Hausaland, and similar formations on the south side of the Sahara. *Geogr. J.* **124**, 527–533.

Grove, A. T. 1969 Landforms and climatic change in the Kalahari and Ngamiland. *Geogr. J.* **135**, 191–212.

Grove, A. T. 1972 The dissolved and solid load carried by some West African rivers; Senegal, Niger, Benue and Shari. *J. Hydrol.* **16**, 277–300.

Grove, A. T. & Warren, A. 1968 Quarternary landforms and climate on the south side of the Sahara. *Geogr. J.* **134**, 194–208.

Grove, A. T., Street, F. A. & Goudie, A. S. 1975 Former lake levels and climatic change in the rift valley of Southern Ethiopia. *Geogr. J.* **141**, 177–202.

Hills, E. S. (ed.) 1966 *Arid lands: a geographical appraisal.* London: Methuen.

Leshnik, L. S. & Sontheimer, G-D. (eds) 1975 *Pastoralists and nomads in South Asia.* Wiesbaden: Harrassowitz.

McGinnies, W. G., Goldman, B. J. & Paylore, P. (eds) 1968 *Deserts of the world: an appraisal of research into their physical and biological environments.* Tucson: University of Arizona Press.

Meigs, P. 1953 World distribution of arid and semi-arid homoclimates. In *Reviews of research on arid zone hydrology,* UNESCO, Paris. Arid Zone Programme 1, 203–210.

Meigs, P. (ed.) 1966 *Geography of coastal deserts.* Arid Zone Research. Review of Research 28 UNESCO.

Monod, T. (ed.) 1975 *Pastoralism in tropical Africa.* Oxford University Press.

Musk, L. F. 1976 Rainfall variability and the Walker Cell in the equatorial Pacific Ocean. *Weather* **31**, 34–47.

Pias, J. 1970 Les formations sédimentaires Tertiaires et Quaternaires de la cuvette Tchadienne et les sols qui en derivent. *Mem. O.R.S.T.O.M.* no. 43. Paris.

Stamp, L. D. (ed.) 1961 *A history of land use in arid regions.* Arid Zone Research. Review of Research 17, UNESCO.

Street, F. A. & Grove, A. T. 1976 Late Quaternary lake level fluctuations in Africa: environmental and climatic implications. *Nature, Lond.* **261**, 385–90.

Thornthwaite, C. W. 1948 An approach toward a rational classification of climate. *Geogr. Rev.* **38**, 55–94.

Tricart, J. 1975 Existence de périodes sèches au Quaternaire en Amazonie et dans les régions voisines. *Rev. Géogmorph. Dynam.* **23**, 145–158.
U.N. Economic and Social Council 1975 Obstacles to development of arid and semi-arid zones. Report of *ad hoc* Interagency Task Force, E/C. 8/WG. 1/3.
Williams, M. A. J. 1975 Late Pleistocene tropical aridity synchronous in both hemispheres. *Nature, Lond.* **253**, 617–618.

Discussion

M. R. Miles (*Meteorological Office, Bracknell, Berkshire*). Is it possible that arid and semi-arid lands may get wetter. No one would deny the possibility but when he links it with the increase of CO_2 due to burning of fossil fuel and increased aerosol loading then it becomes a probability that can be the subject of comment. The latest estimates of these effects is a warming due to CO_2 of at most about 1 °C by the end of the century and a cooling of about the same amount due to the additional aerosol.

Now the wetter period in the Sahara to which Mr Grove has referred occurred at a time when the ice-limits were still at more southerly latitudes, than they are today and when temperature was recovering from a level some 5–10 °C below present-day values. It is still not known exactly what brought this large temperature change about. I think the view put forward by Prof. Newell that it was due to the release of heat gained by the deep oceanic waters during the thousands of years of glaciation is among the most likely. Thus the Saharan wetness 5–10000 years ago occurred against a very different set of conditions from those now prevailing. The effects of increasing CO_2 and aerosol loading appear unlikely to bring this kind of wetness about.

A. T. Grove. I understand that there is still some uncertainty as to the effects of increased aerosol loading on global temperatures. It would indeed be a happy coincidence if the effects were to balance those of increased CO_2 due to burning fossil fuel, but I would not think that one can be complacent about this. A global increase of temperature of only 1 °C is not negligible in importance; such an increase is unlikely to be uniform over the surface. I do not think we are in a good position to predict the consequences for precipitation of such a warming; it seems to me not unlikely that it could result in increased evaporation, especially from tropical oceans, and some associated increase in rainfall.

E. B. Worthington (*I.B.P., c/o The Linnean Society, Burlington House, Piccadilly, London W1V 0LQ*). Important new information has been presented concerning the changes of climate in Africa, and their dating since the last ice age. The former much greater extent of lakes, for instance in the southern Sahara, and in the Eastern Rift Valley, related closely to the present distribution and evolution of fishes. Conversely the fishes had contributed evidence of past climatic change – a nice example of the relation between two quite different disciplines.

H. Doggett (*Leader, Cereals Improvement, I.C.R.I.S.A.T.* 1–11–256, *Begumpet, Hyderabad* 500016, *India*). An area in which more reliable prediction of rainfall and drought years could be of value is in the determination of cropping patterns. Seed stocks of shorter term varieties, or even of different very short-term crops, could be stored, and issued to farmers, with appropriate advice, when such seasons are expected. This would help to reduce the logistical problems of moving in food supplies, and the size of the stocks needing to be stored against the risk of a succession of drought years. The 'mid-season correction' is being used in India, so that when the early weeks of the rains are poor, farmers are advised to grow short-term crops such as setaria millet and mung bean in the resultant enforced late plantings.

B. Dasgupta (*Institute of Development Studies, University of Sussex*). As one among the few social scientists present in this gathering, let me first thank the speakers for their valuable contribution. My point relates to the suggestion regarding the feasibility of switching the cropping pattern to drought-resistance crops in response to mid-season weather forecasting. This suggestion presupposes the unrestricted flow of information to the farmers, their understanding of the various cropping practices, and the availability of seed stocks of different varieties of crops at the village level. My contention is that, apart from the fact that very few villages possess the requisite seed storage capacity, the low level of literacy, the low development of mass media and given the highly stratified and hierarchically based character of many villages, particularly in South Asia, knowledge is a precious commodity which is least accessible to those who are most vulnerable to weather changes, that is the poor farmers.

B. H. Farmer (*Centre of South Asian Studies, University of Cambridge, Laundress Lane, Cambridge CB2 1SD*). I would like to pursue the point made by Mr Jain about mid-crop weather forecasts and, in particular, Dr Gupta's comment on the difficulty of getting cultivators in the villages to act on such forecasts. First, it cannot be too strongly or too often emphasized that to adjust cropping patterns on a government research station is one thing; but to get village farmers to adjust their patterns is quite another. This applies in general, and in particular to adjustments to mid-crop weather forecasts which indicate, let us say, that there will be an early withdrawal of the Monsoon. Secondly, I would suggest (on the experience of my Centre's research project on the so-called 'Green Revolution' in North Arcot District of Tamil Nadu, South India) that the problem may not so much be one of information: our cultivators were surprisingly well informed on a wide variety of agronomic matters. Much more, I would say, is it a problem of the perception of the cultivator (especially the small cultivator) of the relative risks and likely benefits of leaving his sown crop in the ground in the hope of sufficient rain (despite the forecast) and of ploughing it in and resowing with another crop (always assuming he can obtain the necessary inputs) because the forecast is that the rain will be inadequate for the crop already in the ground.

Phil. Trans. R. Soc. Lond. B. **278**, 477–491 (1977) [477]
Printed in Great Britain

The edge of the desert: the problems of poor and semi-arid lands

BY DHARMA KUMAR
Delhi School of Economics, University of Delhi, India

Malnutrition and undernutrition are very widespread in the poor countries (l.d.cs) including several semi-arid lands.

Between 1969–71 and 1985 the output of food in l.d.c.s is likely to grow a little slower than their populations, according to the F.A.O.; since incomes will also rise, the total demand for food will rise by 3.6 % p.a. Many countries will find considerable difficulties in importing the food required.

There has been an enormous increase in irrigation since 1949, much of it financed by l.d.c. governments or by international aid. The financial returns to Governments on large scale irrigation works have frequently been very low, because of a deliberate policy of subsidizing farmers, mismanagement or neglect of complementary inputs. Also, irrigation works in semi-arid lands are often built as protection against crop failure, not to maximize revenue.

Instability in agricultural output can be dealt with by technology to stabilize output, the improvement of methods of predicting changes in output, and insurance or relief. Population pressure has closed many traditional sources of protection, such as temporary migration, as a result of growing populations and the change to modern agricultural technologies. Rural public works programmes have been advocated as a remedy but have generally suffered from a shortage of productive projects and the high non-labour content of productive works.

Finally, measures of population control are discussed.

Some 40 of the countries containing semi-arid lands are very poor; their average income per head is less than $250 per year. In addition, their distribution of income is frequently more unequal than in rich countries. Consequently, large numbers suffer from abject poverty, in particular from hunger and malnutrition. Populations are growing fast, and may grow faster than food production in some countries, mostly in Africa.

After a summary of trends in population and food production, based on the F.A.O., the paper discusses some of the problems facing the governments of poor countries. Most of these countries are largely agricultural, and a basic task is to raise agricultural output. There has been an enormous investment in irrigation, but frequently with inadequate returns to the economy as a whole or the government. In semi-arid lands there is the additional problem of variability of output, and the paper discusses measures to reduce variability or to compensate for the damage caused by it. The paper then considers two long term problems: the growth of unemployment, and the growth of population.

This paper contains little that is not familiar, often tediously familiar, to economists working on rural development, apart from some facts about India, but may be useful to others as a brief and unsystematic survey. My choice of problems and data is doubtless strongly coloured by the fact that the only country I know about in any detail is India; this means that the paper concentrates on the problems of a densely populated country. If I am parochial, at least my parish is very crowded.

The dimensions of mass hunger

It is impossible to be precise about the numbers of people actually suffering from malnutrition. In the first place, estimates of the average daily requirements of any given population are based on assumptions regarding normal body mass, levels of activity, and the age and sex structure of the population, on all of which there may be differences of opinion. Thus estimates of average Indian requirements range from 2373 kcal (9.85 MJ) and 42.74 g of protein to 1789 kcal (7.49 MJ) and 25.9 g of protein (Ryan, Sheldrake & Yadav 1974). India had less than its minimum *total* requirements in 1969–70 by the first estimate, but not by the second. Secondly, figures of average income and food availability are very imprecise (Burk & Mordecai 1976). Thirdly, since income and consumption are very unequal in most countries, the average figures cannot tell us how many people actually got the minimum. Figures of actual consumption are much more difficult to estimate than requirements, and very few poor countries have the data required.

But it is difficult to disagree with the F.A.O.'s statement that 'Each strand of evidence... is far from incontrovertible, but together they point to a conclusion that the problem of food deficiency is extremely serious: one half of child deaths are in some way attributable to malnutrition; there are at least 200 million living child sufferers; and the percentage of population in the developing world subject to food deficiencies ranges to 30 per cent, or even higher in some areas' (Food and Agriculture Organization 1975). The countries suffering most from mal- and under-nutrition include Angola, Somalia, Tanzania and some of the Sahelian countries in Africa; Bangladesh, India, Indonesia and the Philippines in Asia; Afghanistan, Saudi Arabia and the two Yemens in the near East; Bolivia, El Salvador and Haiti in Latin America (U.N. 1975).

Till recently the developing countries have managed to increase food production more rapidly than population. But there are several disturbing facts about the last 20 years. In the first place, the difference between the two rates of growth has been narrowing; indeed there may have been a decline in food production per capita since 1970. And this decline occurred despite government policies to increase food production. Population figures even for the past are subject to wide margins of error, and forecasts of the future are naturally much more uncertain. The F.A.O. forecasts are gloomy, but expected rates of growth of population may be too high.

The population of what the F.A.O. calls the developing market and I shall call the poor non-Communist countries, is expected to grow by 2.7 % per annum between 1969–71 and 1985; this is indeed the U.N.'s 'medium' projection. The F.A.O. assumes that the rate of growth of food production will remain stable at the past rate of 2.6 %, or a little below the rate of growth of population. The effective demand for food will also rise, as incomes rise. Assuming that incomes per head will continue to grow at the same rate as in the past, the demand for food will rise by another 0.9 %, or by a total of 3.6 % per annum. On these projections, poor non-Communist countries will have a deficit of nearly 85 million tonnes (Mt) in 1985, as against an annual average deficit of 17 Mt in 1969–71, which was itself difficult enough to finance. Indeed if one excludes the cereal exporters, the deficit in a normal year might amount to 100 Mt, and in a bad year to 120 Mt or more; in 1973–4 the cereals deficit (excluding the exporting countries) was 52 Mt. World production is expected to be able to meet this deficit; the developed countries are expected to increase their production by 2.4 %

per annum, and although their per capita consumption will rise, their populations will only increase by 0.9 %, leaving them with substantial surpluses. But not all poor countries will be able to buy the food they need, nor is food aid likely to cover the deficit. In terms of the gap between output and population, the problem is most severe in Africa: between 1969–71 and 1985 its population is expected to grow by 76 %, the output of food by 45 %. In terms of sheer numbers, of course, the Asian problem is the worst.

The economic structure

The policies required to increase food supply – by increasing domestic output of food, or of exports to buy food with, or aid – will vary with each country's economic structure and resources. Most of these countries are still largely agricultural: agriculture typically accounts for between one-third and one-half of the gross domestic product, and provides employment for two-thirds or more of the labour force. There is, of course, considerable variation even among the poor semi-arid countries. In Zaire, agriculture accounts for 8 % of the g.d.p., and in Rhodesia for 16 %, but in India for 45 % and in Gambia for nearly 60 % (table 1).

The proportion of a country which is semi-arid also varies enormously, from 5 % in Pakistan to 100 % in Rhodesia and several other countries of Africa. 54 % of India is semi-arid; some 1.7×10^6 km^2 support 261 million people, or over half the total population of the semi-arid tropics. The semi-arid regions of India (with a man–land ratio of 1.54 people to the hectare) are the most densely populated of all the semi-arid regions. In general, the man–land ratio is much higher in semi-arid Asia (1.33) than in Africa (0.13 south of the Sahara, 0.03 in the Near East and NW Africa) (table 1). There are similar variations in the density of animal populations: in Asia there are 1.36 cattle and buffaloes per hectare of arable land, in Africa south of the Sahara 0.89.

In some of the sparsely populated countries much of the population is nomadic. In Somalia 75 % of the human population is nomadic and in Sudan 30–40 % (Grigg 1974). The proportions are much smaller in other countries but many of them, even countries as densely populated as India, contain large nomadic and semi-nomadic populations.

Where there is settled agriculture, the dominant mode of cultivation is small-scale peasant farming. But the land is very unequally divided: in the states in India containing large proportions of semi-arid lands, over half the households cultivated holdings of less than 1 hectare, and occupied only a little over 5 % of the total area; over half the area was cultivated by less than 10 % of the households, though even they were small enough: holdings over 20 ha are rare. There was little difference, in fact, between these and other states in India. The all-India figures exaggerate the extent of true inequality in that farm size is inversely correlated with proportion of land irrigated; this is probably true of semi-arid states also.

The all-India figures show that 45 % of all farms have some irrigated lands; figures are not separately available for semi-arid areas. Again, in India, even in the smallest size group, up to 0.19 ha, the average farm has 1.63 fragments; the number increases with the size of the farm, and farms over 20 ha contain over 11 fragments on average (Ryan 1974). Fragmentation has obvious disadvantages; it is often wasteful of land and time; it may make the use of certain agricultural techniques, including the full use of irrigation water, difficult. But it also has some less obvious advantages: farmers often like to hold fragments of different types of land to spread risk, spread labour over the year, and for ease in selling and mortgaging small amounts of land (Kumar 1975). In arid lands the risk factor may be especially important, since rainfall is so erratic.

TABLE 1.[a] SELECTED SOCIO-ECONOMIC DATA ON SEMI-ARID TROPICAL COUNTRIES[b]

	portion of country semi-arid	semi-arid area	semi-arid population	semi-arid popn/land ratio	population growth rates[c]	per caput g.d.p.[d]	agric. as percentage of total g.c.p.[c]	net food supplies per caput per day		
								energy		protein
	%	10^3 km²	10^6	people/ha	%	$ U.S.	%	kcal	MJ	g
Asia										
Burma	25	170	7	0.41	n.a.	92	34	2010	8.42	44
India	54	1700[e]	261[e]	1.54	2.2	99	45	1990	8.33	49
Pakistan	5	40	3	0.75	n.a.	136	38	2410	10.2	55
Thailand	60	308	25	0.82	2.7	176	29	2210	9.25	51
sub-total or average	36	2218	296	1.33	2.45	106	37	2013	8.43	49
Africa South of the Sahara										
Angola	60	748	3	0.04	n.a.	201	n.a.	1910	8.0	40
Botswana	100	569	1	0.02	2.8	n.a.	43	n.a.	n.a.	n.a.
Cameroon	25	119	1	0.08	n.a.	162	50	2230	9.33	59
Cent. Af. Rep.	50	311	1	0.03	2.2	131	31	2170	9.08	48
Chad	40	514	3	0.06	2.3	69	48[h]	2240	9.37	78
Dahomey	80	92	2	0.22	2.5	24	36	2170	9.08	52
Ethiopia	50	611	12	0.20	1.8	72	52	1980	8.29	66
Gambia	100	10	1	1.00	2.1	123	59[h]	2320	9.71	62
Ghana	50	119	4	0.34	3.0	260	n.a.	2070	8.67	43
Guinea	70	172	3	0.17	2.2	106	n.a.	2060	8.62	45
Kenya	90	524	10	0.19	3.1	135	31	2200	9.2	68
Madagascar	50	294	4	0.14	n.a.	107	30	2240	9.48	51
Malawi	75	88	3	0.34	n.a.	58	48	2400	10.1	63
Mali	60	721	5	0.07	2.1	77	48[h]	2130	8.92	68
Mauritania	25	258	1	0.04	2.2	211	42[h]	1990	8.33	73
Mozambique	50	376[e]	4[e]	0.11	n.a.	228	n.a.	2130	8.92	40
Namibia	30	247	1	0.04	2.0	n.a.	n.a.	n.a.	n.a.	n.a.
Niger	40	507	4	0.08	2.7	91	51	2170	9.08	78
Nigeria	75	693[f]	42[f]	0.61	2.5	97	49	2290	9.58	50
Portuguese Guinea	100	36	1	0.28	0.9	n.a.	n.a.	n.a.	n.a.	n.a.
Rhodesia	100	389	5	0.13	3.3[g]	225	16	2550	11.4	73
Senegal	100	196	4	0.20	2.4	209	33[h]	2300	9.62	64
Somalia	50	319	2	0.06	2.3	70	n.a.	1770	7.41	57
Sudan	60	1491[e]	15[e]	0.10	2.8	137	32	2090	8.75	59
Tanzania	60	752	10	0.13	2.6	77	34	1700	7.12	43
Togo	80	45	2	0.44	2.5	138	43	2210	9.25	51
Upper Volta	100	274	5	0.18	2.1	46	44	2060	8.62	70
Zaire	10	234	2	0.09	4.2	n.a.	8	1040	4.35	33
Zambia	100	753	4	0.05	2.9	399	6	2250	9.41	69
sub-total or average	65	11489	157	0.13	2.5	123	38	2155	9.02	60
Southern and Central America										
Bolivia	20	220	1	0.05	2.6	197	19	1760	7.37	46
Brazil	12	1036	20	0.19	2.8	422	13	2820	13.92	67
		(840)[i]	(14)[i]	(0.16)[i]		(150)[j]				
Paraguay	60	244	1	0.04	n.a.	250	33	2540	11.3	65
sub-total or average	31	3326	58	0.17	2.7	469	19	2670	12.5	67
Near East and Northwest Africa										
Arab Rep. Yemen	20	39	1	0.26	2.7	61	n.a.	1910	8.0	57
Peoples Rep. Yemen	10	29	–	0.17	3.0	113	n.a.	2020	8.45	67
sub-total or average	12	611	2	0.03	2.9	303	13	1997	8.36	58
Oceania										
sub-total or average	34	19565	512	0.26	2.5	159	29	2132	8.42	54

In many dry farming districts, markets are very undeveloped. But debt is common. In their survey of West Rajasthan, Bose, Malhotra, Johory & Sen (1965) found that while the cultivators could finance the kharif crop themselves, they borrowed for the rabi or winter crop, repaying after the harvest 1.25 kg in grain for every kilogram taken.

Those designing Government policy or scientific research must bear in mind that they are dealing with very large numbers of small farmers; a large proportion of these do not own sufficient land to occupy them for more than six months of the year or to feed their families satisfactorily, especially in bad crop years. Again, although the number of landless labourers is generally lower in semi-arid areas than in the more densely populated humid areas, there still are several millions (Sundaram 1975).

Increasing agricultural production

Most Governments rely largely on irrigation to increase agricultural output in the semi-arid zones – and in many countries, irrigation in these areas is the condition for any expansion in the total cultivable area. There are various problems in interpreting international data but even so it is certain that there has been an enormous increase in irrigation since the Second World War: in 1949 the total irrigated acreage in the world was 107.7×10^6 ha; in 1964 the irrigated arable lands alone amounted to 175×10^6 ha (Baumann 1975). A large part of the new irrigation projects has been financed by Governments; in India about half the total irrigation is provided by Government. In particular, a substantial percentage of the revenues of developing countries and of international aid has gone into large scale irrigation projects. But there seems to be a growing feeling that many of these investment decisions were mistaken.

Much of the criticism of large scale irrigation appears to be based on evidence that the return on existing schemes is low. Wollman (1970) has asserted, but without providing evidence, that 'For most of the countries in the arid zones of the world, the scattered and incomplete evidence that is available indicates that output could be doubled or tripled with the same water base if better farming methods were adopted'.

There are two separate questions: the returns to the economy as a whole, and the returns to Government on public irrigation works. Before independence, water rates for Government irrigation works contributed to public revenues in India; since independence there has been a steadily mounting loss (Ministry of Irrigation and Power 1972). This is partly because the rates were very low – a fraction of what farmers were prepared to pay private owners of

NOTES TO TABLE 1

[a] Taken from Ryan (1974); Ryan gives data for all countries, but here countries with a per caput g.d.p. above $250 have been excluded, except for Ghana which is a marginal case. Most of the data relate to the late 1960s.

[b] Semi-arid areas and populations were estimated by using Troll's (1966) map for climatic regions of the world by estimating the regions V_3 and V_4 (dry savanna and thorne savanna) in each country and examining population densities in these areas as contained in *Hamond Incorporated Atlas* (1972). n.a. denotes data not available.

[c] Source: United Nations (1973).

[d] Source: Food and Agriculture Organization (1970).

[e] Excludes the areas irrigated from major government irrigation schemes.

[f] The 1973 Census in Nigeria puts the population of Northern Nigeria at 51 million, the area at 730000 km and the population/land ratio at 0.7. Source: Kassam, Dagg, Kowal & Khadr (1974).

[g] Southern Rhodesia only.

[h] Source: Curtin (1974).

[i] Figures in parentheses are estimates of Paiva, Schattan & de Froitas (1973).

[j] Estimated g.d.p. per caput for northeastern Brazil as contained in Schuh (1973, p. 4).

tube-wells. (There are legal provisions for the collection of betterment levies, but in most States these are ignored.) There is no justification for such low rates which amount to a subsidy to a minority of farmers, frequently the better-off. Raising the rates would not decrease water use (Ansari 1968). In fact keeping rates so low that they do not even cover maintenance costs leads to waste and over-watering both in India and Australia (Clark 1970). But several Indian States have raised water rates steeply in recent years.

On the other hand, even with low rates, the water from large irrigation schemes is sometimes underutilized. Data on utilization of irrigation have to be interpreted with care, particularly where the water is also used for power. It takes several years and considerable investment in feeder channels, ground levelling, etc., to make full use of irrigation facilities. Marketing facilities for inputs such as fertilizers and for agricultural produce may have to be developed. If farmers are used only to dry farming, it may take additional time to switch to wet farming, which needs more labour and capital per hectare. So in a period of rapid expansion of irrigation, such as the last 30 years, statistics of underutilization may reflect largely the problems of adjustment. (In fact the intensity of utilization has risen from about 60 % at the end of the First Plan, in March 1956, to 83 % by March 1972, Ministry of Irrigation and Power 1973.)

But there are other problems which will not be cured merely by the passage of time. Waterworks may be mismanaged (Wade 1975). Farmers buy fertilizers and other inputs, and the only source of credit may be the village moneylender or rich farmer, whose own resources are limited, and whose rates of interest are high. The fragmentation of holdings may make it uneconomic for an individual to build feeder channels; the social structure of the village may not encourage cooperation. Sometimes even large farmers do not make full use of irrigation facilities: it may pay them to rent out the land to others willing to undertake the necessary expense and labour, but they are afraid of losing the land permanently under present land laws.

In the belief that farmers on their own cannot make full use of irrigation facilities, the Government of India has assumed the responsibility not only for on-farm development, including the provision of credit, but also for building roads, marketing, providing inputs and extension services. The Command Area Development Programme is only a few years old; only a few C.A.D. authorities have actually started working, management problems are still unsolved, and it is too early to judge the success of the experiment. The Government of India is even considering a bill which would empower State Governments to regulate the kind of crops farmers can grow and the period of sowing on land irrigated by Government: if this involves detailed administration, experience suggests that it is unlikely to be successful.

The financial returns on Government irrigation works in India have been low, first because the price charged for the water has been low, diverting a part of the benefits of the investment in irrigation to the farmers. This is true of many other developing countries (Duane 1975). Secondly, the return to the direct users of the water has also been below potential because of mismanagement in the distribution of water, lack of infrastructure, etc. There is a third important reason for the low social returns which is to some extent a matter of deliberate choice. Irrigation works have been built in semi-arid areas as a measure of protection against failure of crops. If the Government could concentrate only on maximizing total output it might locate irrigation projects differently, but it has to consider problems of regional balance.

Protection against uncertainty

The variability of output is one of the main problems of rain-fed agriculture. The semi-arid regions of northern Africa and western Asia are particularly prone to very sharp declines in output. In India too the dry areas are the major source of instability. The main reason is rainfall: the coefficient of variation of rainfall is inversely correlated with mean rainfall.

The F.A.O. (1975) has calculated 'instability indexes' of food production for 125 countries, or the coefficient of variation from the trend rate. Comparing the indices for the two periods 1952–62 and 1962–72, the F.A.O. found no significant trend; for the world as a whole the instability index was 4. But the overall index for developed countries was 6, whereas that for the l.d.cs was 3; the output of the d.cs is thus twice as unstable as that of the l.d.cs. To some extent this may have been brought about by Government policies, such as support schemes for agriculture. But the literature is full of references to the uncertainty of l.d.c. agriculture; presumably in order to stress the fact that its variations are much more uncontrolled than in the developed countries, and also that they are much harder to bear in l.d.cs.

I have not been able to obtain the F.A.O. calculations for each country (assuming that they were made for each country) but it is probable that the output of the semi-arid regions is more unstable than in the rest of the developing world.

The lack of markets increases instability. Thus the Indian farmer grows bajra, because he needs cereals, even though it is less stable than alternatives, such as kharif pulses (Jodha & Vyas 1969).

There are three lines of attack on this problem. First, one can try to alter the techniques of production so as to make output less variable and less uncertain, by irrigation, mixed farming and research into drought resistant varieties. In the past, scientific research has often neglected those crops most useful to poor countries, but this imbalance is being rectified; research is in progress on the cereal grains and the grain legumes of the semi-arid lands. The latter are the cheapest source of protein even at the very low current yields.

Secondly, one can improve one's method of predicting changes in output. Agrometeorology is still in its infancy; also, meteorologists, like all other valuable resources, are very unequally distributed (F.A.O. 1975). International action such as the proposed world agrometeorological watch should reduce uncertainty.

Thirdly, where risk is unavoidable, one can devise ways of reducing its impact on those who cannot or should not bear it. In India famine relief has always been one of the responsibilities of the Government, and under the British an elaborate system of famine relief evolved, ranging from the remission of land revenue to the setting up of public works and relief kitchens. Since independence, the expenditure on relief has grown enormously. Since relief was genuinely urgently required it was sanctioned hastily and actual expenditure was not sufficiently scrutinized. 'It would perhaps be no exaggeration to state', the Planning Commission Report on Integrated Agricultural Development in Drought Prone Areas remarked, 'that there is no other activity in which demands are processed as perfunctorily, funds of large dimensions allocated so easily and expenditure is incurred with such inadequate planning and scrutiny as in the case of relief operations' (Planning Commission 1973).

Several writers have described the waste and corruption of the Indian drought relief programmes (Jodha 1975), but as usual suggestions for improved systems are rare. Indeed it seems to have been the experience of these recent drought prone areas programmes that led Morris

(1974) to suggest that relief is frequently not necessary. He argues that the population of semi-arid areas 'have worked out mechanisms for coping with climatic instability'; these include the building up of inventories in good years, which can be drawn down in lean years, and migration when the rains fail. 'Why in a modern nation should regions susceptible to drought suffer more than areas with less volatile weather cycles? The answer is that they probably do not; they only suffer differently.'

There is an important truth in this statement, that the inhabitants of drought prone areas are not necessarily the poorest in the country, even in years of drought, whereas in normal years they may be better off than the inhabitants of densely populated areas (Chatterjee, Sarkar & Paul 1974). The Governments of poor countries are poor, and have to make hard choices between poverty relief programmes for different regions and groups. The survivors of dramatic disasters are frequently overcompensated, though less so in poor countries than in rich. But this still leaves a number of questions unanswered.

In what sense is India 'modern' – after all, part of the problem is precisely that transport and markets are underdeveloped. Were traditional methods of coping with instability ever particularly 'efficient', however one defines efficiency? They have not prevented deaths in the past. Nor is famine a good agent of natural selection of farmers. Since inhabitants of semi-arid lands *do* suffer differently, special measures may be needed. Is instability of output increasing as a result of population pressure? Can traditional measures continue to be used?

Jodha (1975) has described in more detail the measures with which Indian farmers traditionally meet droughts and famines. As soon as a crop failure is expected, labour is used more intensively in cleaning the fields and weeding, collecting rough fodder, processing chaff before feeding it to the animals and so forth. Once the crop failure occurs farm families immediately restrict their consumption. Inventories are run down and assets mortgaged; in extreme distress jewellery, land and cattle are sold. Farmers migrate from drought stricken areas as a last resort. Sometimes only a man or two from each family moves to nearby towns or irrigated areas in search of temporary employment, sometimes whole families with their livestock may travel several hundred kilometres in search of employment for human beings and pasture for animals. Some indication of the extent of migration is provided by the fact that three million inhabitants of the arid zone in India alone migrate every third year in India (Planning Commission 1973).

But no modern government should be satisfied with these traditional responses to scarcity. The sharp curtailment of consumption leads to the enfeeblement and deaths of humans, especially infants, and animals. Farmers who sell their land and cattle during famines, when prices are low, find it difficult to buy them back in normal times, when prices have risen: Jodha has reported that the droughts have increased inequalities in rural areas. But it is not clear to what extent this is cumulative; as we have seen, land is generally not more unequally distributed in dry areas than in wet.

The pressure of population is closing many of the traditional sources of protection against scarcity. The traditional system of rotational grazing is being abandoned, pastures are being ploughed up, and sub-marginal lands are being cultivated. In these areas crop yields have fallen, so that it is difficult to stock food and fodder even in good years. Again, while the need to migrate has increased, farmers have to go further, and Jodha reports that since 1956–7 or so, the councils of some villages to which the people of Rajasthan traditionally migrate charge increasingly heavy taxes. (Compare this with the problems of international boundaries faced by migrants in the Sahel.) Also, the migration of cattle often involves over-grazing: since the owners of the herds have little interest in preserving the pastures there has been extensive destruction of natural pastures along traditional routes in India. There is clearly need for a well-worked out policy of famine or drought relief. The question of relief works overlaps with the wider problem of maintaining employment.

Rural public works

In most l.d.cs there is already a large amount of surplus labour, if one includes besides the openly unemployed those who work only part of the day. In India, nearly 70 % of the work force is in agriculture, subsistence farming predominates; open unemployment is small, and estimates of disguised unemployment vary. Mehra (1966) estimated that in 1956–7 about one-third of the workforce in agriculture could be removed without reducing output. Under-employment was not closely correlated with climatic conditions: some semi-arid states, such as Rajasthan, Punjab and Madhya Pradesh had high rates of disguised unemployment; Gujarat, Andhra Pradesh and Maharashtra not (Mehra 1966). Unemployment has almost certainly declined in Punjab and Haryana, but trends in the rest of India are not clear (Sundaram 1975). Other semi-arid tropical countries also suffer from open or disguised unemployment, 60–70 % of the rural work force in the Maghreb according to one estimate (Tiano 1973).

Since the population of l.d.cs has been growing rapidly over the last 15 years, the labour force over the next 15 years at least is bound to grow too. An estimate made in 1971 predicted that the labour force would increase by 2.3 % between 1970 and 1980 (Turnham 1971); in many l.d.cs the proportion of the population that is openly unemployed will increase.

The scale of the problem of unemployment is unprecedented. Neither in Europe nor in Japan did the labour force grow by over 1.6 % per year during the period of development; the United States experienced a rapid growth in the labour force between 1850 and 1914 but it had resources to match. It is worth stressing this point since some economists argue that the l.d.cs of today should learn from the advanced countries who relied on labour-saving technology during their period of industrialization (Robinson 1975).

The population in the rural areas is growing rapidly in most l.d.cs; if rural employment could increase at the same pace, the costs of urbanization would be saved. But there is a growing disquiet in the literature on the effects of introducing modern agricultural technology on employment, both in agriculture and in the economy as a whole (Raj Krishna 1974; Singh & Day 1975). To illustrate, Raj Krishna has calculated both the direct and indirect effects on employment of a 5 % growth in output, with technological change of the type experienced in wheat cultivation in the Indian Punjab between 1968/9 and 1973/4. He finds that with technological change total employment will only grow by 2 % because agricultural employment grows very slowly, mainly as a result of the mechanization of threshing and the shift from bullock driven to power driven irrigation. Without technical change a 5 % increase in agricultural output increases total employment by 4.7 % (the increase in non-agricultural employment is much smaller in the second case). Unless further, more detailed, calculations alter these findings very sharply, Governments must consider the possibility that the growth of certain types of 'modern' agriculture will stimulate growth in output and employment in the rest of the economy, but at a rate insufficient to absorb the increase in labour force that is likely.

If spontaneous developments in agriculture cannot provide the additional employment required, can Government do so? The real issue seems to be not whether it should or should not, but whether it can. One can argue that it has in its favour resources of money, organizational skills and technology; or that it alone can remove certain obstacles to agricultural development – it can enforce consolidation of fragmented holdings or redistribute land. Also, the rapid growth of agricultural output should make it possible to operate public works on a large scale; workers can be paid indirectly out of the increase in food supply. Unfortunately, the experience in India and other countries with public works schemes, such as Algeria, Morocco and Tunisia, has so far not been happy (Lewis 1972; Tiano 1973). The evaluation reports submitted to the Government of India and numerous reports make the same accusations: bad planning, inefficient implementation, red tape, waste and corruption. Many roads and earthworks were washed away by the first rains while contractors made huge profits. But the search for ways of improving public works schemes rightly continues.

An interesting experiment in new institutional structures for increasing rural employment is the Labour-cum-Development Bank set up by the Government of Kerala. This Bank finances small quick yielding labour intensive rural projects; the beneficiaries have to repay the loan within three years. A special feature of the scheme is that labourers are paid two-thirds of the market wage in cash, the balance is withdrawable after three years, with 9 % interest; an anti-inflationary measure. The Bank has only been in operation since 1973, but it has already found a shortage of properly processed schemes; also, while it was originally planned that the labour content of the projects would be 80 % of the total cost, in the three schemes it has financed so far, wages amount to less than 60 % of the cost (State Planning Board, Kerala 1974). These have been the recurrent problems of nearly all rural works programmes in India: the shortage of productive schemes ready for implementation, and the high non-labour content of those schemes which are productive.

There is clearly no shortage in theory of socially productive public works: irrigation channels, drainage, roads, village schools and hospitals, storage facilities, housing, the list is endless (Lewis 1972). (Many of the semi-arid districts of India were once centres of great civilizations and the repair of the monuments they have left behind them would be very useful, even from the narrow viewpoint of tourism.) The problem is the cost of the non-labour inputs, a good problem for technology to solve, especially since the provision of employment ranks high amongst national objectives in India and elsewhere. The State of Maharashtra already operates an employment guarantee scheme under which in every district manual work is to be provided to all who want it, if possible within 5 km of their homes. Wages are based on output at a rate enabling the average worker to earn Rs. 3 for a seven-hour day, a rate which compares favourably with the current daily wages of field labourers. Nearly 80 % of the expenditure incurred so far has been on irrigation works (Planning Department, Government of Maharashtra 1975) as in nearly all the other employment schemes in India. Unfortunately no study has been published of this scheme, which has been in operation since 1972.

Population growth

In many developing countries and certainly in Asia, a reduction of the rate of growth of population would ease the problems of unemployment and low economic growth, and it is clearly preferable to achieve this by lowering the birth rate, and by voluntary means. Nearly all poor societies already practice various traditional forms of birth control (Cowgill 1975;

Mandelbaum 1974) though their efficiency can doubtless be improved upon. There are two types of actions Government can take to encourage a decline in fertility. There are direct family planning programmes providing information, free contraceptives and positive incentives (from tax relief to free radios and subsidized houses) and disincentives (such as jail or compulsory sterilization). Indirect measures would work upon the motives for having large families, such as support in old age.

The influences on fertility are still imperfectly understood (Schultz 1976). Economists have put forward two main economic reasons for large families in poor rural societies. One is that after the first five years or so, children can add to the family income, by doing work on the family farm, cattle herding and so on. This is the aspect Robinson stresses:

'One very clear reinforcer for high fertility is the traditional rural family life-style with its great need for family labour.... To sum up, the traditional peasant family unit is labour intensive and of high fertility and the two characteristics interact and reinforce one another. Any policy which encourages labour-intensive techniques may also be unwittingly striking a serious blow at any hope of getting a small family accepted' (Robinson 1975).

But he ignores the second economic reason: the need for security in old age. It is likely that fertility is also determined by the need to ensure that at least one son will survive to support the family (Mandelbaum 1974).

Public schemes for support in old age might therefore reduce the demand for sons. One could even argue that widespread and prolonged unemployment would increase fertility, if families took as given the probability distribution of employment, but these speculations can hardly be tested. And social security measures in India, such as the Tamil Nadu old age pensions scheme, or the Maharashtra Employment Guarantee Scheme, have not lasted long enough to have significant effects on fertility.

There is evidence both from Pakistan and Bangladesh that an increase in female labour force participation leads to a decline in fertility; in Bangladesh this was true of women working in agriculture (Chaudhury 1974; Shah 1975). As far as I know, no one has discussed the effect of modern agriculture on fertility via female participation rates – for instance weeding is a largely female occupation but herbicides supersede human weeders. Bardhan (1974) has speculated on the effects different systems of agriculture have on death rates: female children may be more neglected in arid hilly areas than in the irrigated areas, especially those growing rice, where women are needed for weeding, harvesting, and threshing.

If he is right, female children will be neglected in the plains too, as female labour is displaced!

Apart from its intrinsic interest, a deeper understanding of the motives for bearing children would improve the efficiency of family planning programmes and the effects of other measures increasing employment, income and social security, on family planning. But one need not, of course, wait for large changes in the economy: a survey of rural communities in Maharashtra during 1953–66 found that the birth rate had declined, even without significant improvements in income, education or health facilities, as a result of a vigorous family planning campaign, in particular sterilization camps (Dandakar & Vaijayanti 1975).

Several developing countries have already succeeded in reducing their birth rates significantly: American Samoa, Chile, Costa Rica, Egypt (from 43 in 1960–4 to 35 in 1971), Fiji, Guadeloupe, Jamaica, Martinique, Mauritius, Puerto Rico, Reunion, Sri Lanka, Trinidad and Tobago and Tunisia (from 46 to 35). Fertility has also declined in parts of Brazil (U.N. 1974). In contrast, in some countries of Africa fertility, already high, may have risen as the incidence of debilitating diseases has fallen. But the expectation of life in Tropical Africa is still the lowest in the world: between 41.3 (in West Africa) and 44.5 (in East Africa). It is hopeful too that measures which seem to be most important in bringing about a fall in fertility are good in themselves; better health facilities, better education, especially female education, more social security.

I am grateful for their comments to several of my colleagues at the Delhi School of Economics, and in particular M. Datta Chaudhuri, J. Krishnamurty and K. Sundaram, as also to Robert Cassen, A. R. Jolly, Ashok Khosla and Deepak Lal, K. S. Moorthy, Amartya Sen and T. N. Srinivasan. I am indebted for various facilities to the Institute of Development Studies, Sussex.

References (Kumar)

Ansari, N. 1968 *Economics of irrigation rates, a study in Punjab and Uttar Pradesh.* Bombay: Asia Publishing House.
Bardhan, P. K. 1974 On life and death questions. *Econ. Polit. Weekly* **2**, 1293–1305.
Baumann, H. 1975 Irrigation development in semi-arid areas. *Appl. Sci. Devel.* **5**, 7–2.
Bose, A. B., Malhotra, S. P., Johory, C. S. K. & Sen, M. L. A. 1965 Agriculture in central and lower Luni Basin in Western Rajasthan. *Ann. Arid Zone* **4**, 195–206.
Burk, Marguerite C. & Mordecai, E. 1967 Food and nutrition in developing economies. In *Agricultural development and economic growth* (eds Herman M. Southworth & Bruce F. Johnston). Ithaca, New York: Cornell University Press.
Chatterjee, G. S., Sarkar, D. & Paul, G. 1974 A note on the variation in the dietary levels of households in rural India'. In *Poverty and income distribution in India* (eds T. N. Srinivasan & P. K. Bardhan). Calcutta: Statistical Publishing Society.
Chaudhury, R. Huda 1974 Labour force status and fertility. *Bangladesh Devel. Stud.* **2**, 819–838.
Clark, C. 1970 *The economics of irrigation*, 2nd edn. Oxford: Pergamon Press.
Cowgill, G. L. 1975 On causes and consequences of ancient and modern population changes. *Am. Anthrop.* **77**, 505–525.
Curtin, T. 1974 Africa and the European community. In *Africa south of the Sahara* 1974. London: Europa Publications Ltd.
Dandekar, K. & Vaijayanti, B. 1975 Socio-economic change during three five year plans. *Artha Vijnana* **17**, 305–453.
Dregne, H. E. 1970 (ed.). *Arid lands in transition*, Washington, D.C.: American Association for the Advancement of Science, Publication no. 90.
Duane, P. 1975 A policy framework for irrigation water charges. World Bank Staff Working Paper no. 218, Washington.
Food & Agricultural Organization 1970 *Provisional indicative world plan for agricultural development*, vols. 1–3, Rome.
Food & Agriculture Organization 1975 *The state of food and agriculture* 1974. Rome.
Gopalan, C., Balasubramanian, S. C., Rama Sastri, B. V. & Rao, K. Visweswara 1969 *Diet atlas of India.* India: National Institute of Nutrition, Indian Council of Medical Research, Hyderabad.
Government of India, Ministry of Irrigation and Power 1973 *Report of the Committee of Ministers on Underutilization of Created Irrigation Potential*, New Delhi.
Grigg, D. B. 1974 *The agricultural systems of the world. An evolutionary approach.* Cambridge University Press.
Hamond Incorporated 1972 *Hamond citation world atlas.* Maplewood, New Jersey: Hamond Incorporated.
Jodha, N. S. 1966 The scarcity – oriented growth pattern of aid agriculture. *Ind. J. Agric. Econ.* **21**, 238–248.
Jodha, N. S. & Vyas, V. S. 1969 *Conditions of stability and growth in arid agriculture.* Agro-Economic Research Centre, Vallabh Vidyanagar (Gujarat).
Jodha, N. S. 1975 Famine and famine policies: some empirical evidence. *Econ. Polit. Weekly* **10**, 1609–1624.
Kassam, A. H., Dagg, M., Kowal, J. & Khadr, F. 1974 Food crops in the Sudan Savanna Zone of Northern Nigeria. Unpublished Paper, Institute for Agricultural Research, Samaru, Nigeria.
Krishna, Raj 1975 Measurement of the direct and indirect employment effects of agricultural growth with technical change. In *Agriculture in development theory* (ed. Reynolds).
Kumar, Dharma 1975 Landownership and Inequality in Madras Presidency, 1853–54 to 1946–47. *Ind. Econ. Social Hist. Rev.* **12**, 229–261.
Lewis, J. P. 1972 The public works approach to low-end poverty problems: the new potentialities of an old answer. *J. Devel. Planning* **5**, 85–114.
Mandelbaum, D. G. 1974 *Human fertility in India.* University of California Press.
Mehra, Shakuntala 1966 Surplus labour in Indian agriculture. *Ind. Econ. Rev.* N.S. **1**.

Ministry of Irrigation and Power 1972 *Report of the Irrigation Commission*, vol. 1, New Delhi.
Morris, D. 1974 What is a famine? *Econ. Polit. Weekly* **9**, 1855–1864.
Paiva, R. M., Schattan, S. & de Froitas, C. F. Trench 1973 *Brazil's agricultural sector: economic behaviour problems and possibilities*, XV. International Conference of Agricultural Economics, São Paulo, Brazil.
Planning Commission, Government of India 1973 Integrated agricultural development in drought prone areas: report by the task force on integrated rural development. Mimeo.
Planning Department, Government of Maharashtra 1975 *Employment guarantee scheme, Maharashtra State*. Bombay: Government Central Press.
Reynolds, L. G. 1975 *Agriculture in development theory*. Yale University Press.
Robinson, W. C. 1975 Some negative thoughts about labour-intensive development. *Malayan Econ. Rev.* **20**, 12–26.
Ryan, J. G. 1974 Socio-economic aspects of agricultural development in the semi-arid tropics. Occasional paper 6, Economics Department, ICRISAT, Hyderabad, India.
Ryan, J. G., Sheldrake, R. & Yadav, Satya Pal 1974 Human nutritional needs and crop breeding objectives in the semi-arid tropics. Occasional paper 4, Economics and Statistics Unit, ICRISAT, Hyderabad, India.
Schuh, G. E. 1973 The modernization of Brazilian agriculture. Paper prepared for the U.S. National Academy of Science. Draft mimeo.
Schultz, T. Paul 1976 Determinants of fertility: a micro-economic model of choice. In *Economic factors in population growth* (ed. Ansley J. Coale). London: Macmillan.
Shah, Nasra M. 1975 Female labour force participation and fertility desires in Pakistan: an empirical investigation. *Pakistan Devel. Rev.* **14**, 185–206.
Singh, Inderjit & Day, R. H. 1975 Factor utilization and substitution in economic development: a green revolution case study. *J. Devel. Stud.* **2**, 155–177.
State Planning Board, Kerala 1974 A report on the working of the Labour-cum-Development Bank. *Indian J. Labour Econ.* **16**, 320–349.
Sukhatme, P. V. 1965 *Feeding India's growing millions*. Bombay: Asia Publishing House.
Sundaram, K. 1975 The changing structure of the rural work force in India: 1951–1971. Delhi School of Economics. (Unpublished.)
Tiano, A. 1973 Human resources investment and employment policy in the Maghreb. In *Employment in Africa*. Geneva: I.L.O.
Troll, C. 1966 Seasonal climates of the Earth. The seasonal course of natural phenomena in the different climatic zones of the Earth. In *World maps of climatology* (eds M. E. Lindsberg *et al.*). Berlin, Heidelberg, York: Springer-Verlag.
Turnham, D. 1971 *The employment problem in less developed countries*. Paris: O.E.C.D.
United Nations 1973 *Statistical yearbook* 1972. New York: Statistical Office of the United Nations, 24th Issue.
United Nations 1975 *Report on the world social situation* 1974. New York.
Wade, R. 1975 Water to the fields: India's changing strategy. *South Asian Rev.* **8**, 301–322.
Wollman, N. 1970 Economics of land and water use, in Dregne 1970.

Discussion

G. MURDOCH (*Land Resources Division, Ministry of Overseas Development, Tolworth Tower, Surbiton, Surrey*). I would like to comment on the advantages of fragmentation of cultivated holdings. The risk that is spread relates less to atmospheric moisture distribution than to soil moisture contents and seasonal water-table levels. In years of high rainfall a farmer's upland fields do well though crops on his patches of hydromorphic soil are waterlogged. When rains fail and his well-drained soil yields poorly, adjacent low-lying sites can be relied on reasonably often to produce rations sufficient to avert total famine. In the Swaziland Lowveld a distinctive name – *tihubodla* – is given to arable patches in valleys where relatively high watertables facilitate 'insurance' cropping of this kind.

J. SWIFT (*I.S.I.O., Sussex University, Falmer, Brighton*). I take issue with Kumar's support for a policy of benign neglect in drought and famine areas, a policy which is based on the assumption that peasants and pastoralists have their own adaptive responses to such conditions and can cope better on their own.

There is of course some truth in this assumption; the traditional reactions of peasants and pastoralists to drought and famine did give them some protection. But these responses have long ago been modified in ways that are impossible to reverse, even were this desirable. Indeed it is becoming increasingly clear that famine conditions are at least partly man-made, a result of the economic and political relationships between peasant and pastoral societies on the one hand, and a wider market economy and a central political power on the other.

The Sahelian and Ethiopian famines provide an example. In these regions there has been increasing integration of traditional peasant and pastoral economies to the market economy, resulting in declining food production for own consumption, and increasing production of cash crops for marketing; combined with declining terms of trade between agricultural and pastoral products on the one hand, and manufactured goods on the other, this has led to increased susceptibility to famine in such societies.

In these conditions a policy of benign neglect with regard to famine relief would not mitigate these long-term economic trends.

At the same time, the increasing commercialization of subsistence economies, and the increasingly centralized bureaucratic political power which has replaced previously diffused political power, has progressively deprived people of responsibility for and control over their own land. So an increased susceptibility to famine has been accompanied by increased environmental destruction, which in turn magnifies the famine risk.

D. Kumar. Swift's first paragraph stating that I support a policy of 'benign neglect' is completely baseless; see the section on protection against uncertainty pp. 483–485, especially the concluding sentences: 'There is clearly need for a well-worked out policy of famine or drought relief. The question of relief works overlaps with the wider problem of maintaining employment.'

His other comments raise issues beyond the scope of the paper, which dealt with questions of Government policy. The effects of commercialization and integration into a market economy require much more analysis than is possible here.

N. E. Reynolds (*The Ford Foundation, 55 Lodi Estate, New Delhi 110 003, India*). There have been a number of references to agricultural research, most in an expectant vein. Some speakers have praised the work being done at I.C.R.I.S.A.T. I am familiar with the latter and in no way wish to detract from the compliments paid to it. Nonetheless, there is a danger that these expectations will be disappointed in practice and the reasons for this being so should be recognized. On a popular level there is an expectation that agricultural breakthrough will occur in an orderly progression; from wheat, to rice, to cotton and then, due about now, to the coarse grains, oilseeds and grams that form the bulk of crop agriculture in semi-arid lands. Breakthroughs have and will occur. What must be realized is that the environment for research on and for its adaptation by farmers in the semi-arid areas is much more difficult. Firstly, there is the underlying factor that much, if not nearly all, farming is of a subsistence nature which alters farmers' perceptions from those of researchers and policy makers as to the real value of purchased inputs. The risks attendant on crop agriculture are higher in these regions than in the wheat and rice belts. Much of the farming is already on marginal lands due to population pressures, environmental destruction and a resultant more extensive approach. There is, in most areas, a continuing spread of cropping onto technically unsuitable soils with a heightened erosion threat. Parallel to the last point, the common lands which have traditionally provided a vital flow of nutrients, fuel and power (through traction) – the forest and grazing lands – into crop agriculture have suffered in some areas close to the point of no return. As a result the farming systems are becoming increasingly fragile and exploitive of soil wealth. Under such conditions new crop technology, while offering a chance, can only transform agricultural stability and production if the environmental tissues are also tackled – land use, community relations to resources, physical flows of nutrients, fuel and power into crop agriculture.

A further complication is that, at least in countries with a varied agricultural base, agricultural research constituencies will have arisen or will arise first in the other areas – e.g. with wheat, rice, cotton and sugar in India. As poor countries are able to marshal only limited financial and manpower resources, the capacity of agricultural research to handle all the demands placed on it will be limited. Breakthroughs in the crops named above and the growth of strong commercial interests around those crops will produce strong lobbies. Threats to new production gains, as pest and disease, will demand, and most likely get, the attention of the research establishment. Hence long term, well supported research efforts for the coarse grains, oilseeds and grams by national research organizations may suffer from a series of competitive claims. What this scenario suggests is a mixed approach. Research, yes. But also a serious attempt to grapple with the land-use question and the nutrient and other flows from marginal (and usually common or state lands) into crop agriculture. This complex of issues is largely institutional, market and price in nature and is hard to achieve without strong leadership. Alongside the latter, improved seeds (i.e. good seed not necessarily h.y.v.), improved husbandry practices and improved services and marketing, even on a humble level, will help to lay a foundation upon which farmers can accept the newer risks the use of new technology will involve.

Professor Kumar referred to the wastage surrounding relief expenditures. From her paper and from earlier discussion one might claim that 'community' emerges as the key element in the management of resources. One could add that nowhere, to my knowledge, has the idea of a budget for a village or an area (to cover several years and embodying government expenditures under all heads) been tried whereby government could negotiate a contract with the community. Such a procedure would help elicit greater involvement by the community and build the institution of community – contrary to the common effects of popular government. It would also be an educational tool to raise awareness of alternatives, of short- and long-term trade-offs and of the value of sequential decisions and investments. At the same time it should produce a measure by which government's performance can fairly be measured, in the process providing it with advance information as to the rôles it must be able to play. Then a format would exist upon which larger funds could be expended as a relief measure, disciplined by the budget and the negotiated and contracted plan.

Phil. Trans. R. Soc. Lond. B. **278**, 493–505 (1977) [493]
Printed in Great Britain

The evaluation and exploitation of semi-arid lands: Australian experience

By R. A. Perry
Chief, C.S.I.R.O., Division of Land Resources Management, Wembley, Western Australia

Arid and semi-arid lands are areas where rainfall is too low or too unreliable for cropping or sown pastures. Australian arid and semi-arid lands comprise a single vast area extending to the 250 mm mean annual rainfall isohyet in the south and the 750 mm isohyet in the north. The grazeable areas are used for grazing of sheep or cattle under very extensive management, producing wool and meat for export. The industries have a low production per unit area, but are very efficient in terms of production per unit of labour.

Graziers and administrators have had an overoptimistic perception of the land and climate which has led to varying degrees of resource degradation. Applied research into resource inventory, range condition and trend, stock carrying capacities, and droughts is described.

The land and climate of arid and semi-arid Australia are comparable with those of other parts of the world, but socio-economic factors are very different.

Definition

In Australia, arid and semi-arid lands are together synonymous with the 'pastoral zone', and comprise that part of the continent where rainfall is too low or too unreliable for cropping or sown pastures to be economic. This definition is similar to that of Meigs (1953) for arid land and has two important connotations. First, the emphasis is on inadequacy of rainfall for cropping, that is the biological or agricultural effectiveness of rainfall, rather than on an absolute amount of rainfall (Perry 1968, 1970*b*). It therefore includes a whole complex of factors such as seasonality, reliability, rate of evaporation during the rainy season, and soil aspects concerned with the size and duration of the soil water store. The essential feature for cropping is a reliable, unbroken period of available soil water which is long enough to allow the crop to complete its life cycle. In southwestern Australia where rain falls in winter, when evaporation is low, and from a fairly regular progression of low pressure systems at approximately weekly intervals, wheat can be grown economically under a mean annual rainfall as low as 250 mm even on sandy soils with a low water holding capacity. The situation is quite different in tropical Australia where dryland cropping is possible only where mean annual rainfall exceeds 750 mm. This is because the rain falls in summer when evaporation is high and the rain-bearing influences (convectional thunderstorms and tropical cyclones) are erratic. Below 750 mm mean annual rainfall in these areas, there is a high frequency of dry periods of several weeks duration within the rainy season. During such dry periods the soil water store is depleted completely, thus breaking the continuous period of available soil water required for cropping. A similar situation applies in North Africa where the mean annual rainfall at the boundary between arid and non-arid (or cropping and non-cropping) is much lower under the Mediterranean winter rainfall areas north of the Sahara than in summer rainfall areas south of the Sahara. This is a very important aspect; the lack of recognition by people of European origin of the low biological or agricultural effectiveness of rainfall in summer rainfall areas has been the cause of the failure of many developmental schemes in those parts of the world, including tropical Australia.

The second connotation in the definition used in Australia of arid and semi-arid lands stems from the word economic. Although in the definition it qualifies cropping and sown pastures in higher rainfall areas, it is equally applicable to any industries in arid Australia. The implication is of a monetary, market-oriented cultural system which contrasts markedly with the subsistence systems of many of the world's arid lands.

Rainfall in the area is highly variable and droughts of one to several years duration are a normal feature. Management for both financial and biological performance must be in the context of droughts being a normal feature of the environment.

Description

The Australian arid or pastoral zone is a vast continuous area of 5.7×10^6 km^2, over 70 % of the total area of Australia (figure 1). It extends from latitude 16° S to 35° S and longitude 113° E to 148° E and stretches for over 3200 km east to west and for more than 2000 km north to south. It is second only in size to the arid area stretching across North Africa to the Indian subcontinent (Perry 1970*b*).

Topographically it is flat and low, a vast gently undulating plain broken only by isolated low mountain ranges. The lowest point, Lake Eyre, is about 10 m below sea level and the highest, Mt Zeil in central Australia, is 1500 m above sea level. No large rivers, like the Nile, Indus or Colorado, with headwaters in higher rainfall areas, cross it. It contains no perennial rivers, the water courses being dry except after rain. Most of the area either drains through an ephemeral internal drainage system towards Lake Eyre or has only disorganized drainage with short streams petering out on plains not far from their source.

Much of the area is, or is derived from, an ancient Tertiary land surface with consequent highly leached, infertile soils.

Most of the area receives more than half its rainfall in summer but in the south, winter rainfall is more reliable and effective (Perry 1970*a*). Rainfall variability from year to year is high with consequent high variability in plant growth and forage availability. The lowest mean annual rainfall is 125 mm in the Lake Eyre region. From Lake Eyre, mean annual rainfall increases outwards towards the zone boundary which corresponds to the 250 mm isohyet in the south, the 750 mm isohyet in the north, and intermediate isohyets in the east.

At the level of detail required for management purposes the arid and semi-arid part of Australia contains a great diversity of land types varying in climate, slope, soil and vegetation. However, for broad general purposes most of the area consists of the following few types (Christian & Perry 1969; Leigh & Noble 1969; Perry 1970*a*).

Mitchell grass plains

These are treeless, or almost treeless, plains with cracking heavy clay soils. They are characterized by perennial *Astrebla* spp. (Mitchell grasses). Their main occurrence is in a broad arc around the northern and north eastern summer rainfall part of the zone. The dominant grasses are discrete tussocks 0.5 to 1 m high and 10 to 30 cm in diameter. Other perennial grasses include *Eragrostis* spp., *Aristida* spp., and *Dichanthium* spp. Following rain, annual grasses and forbs grow in the interspaces between the perennial tussocks.

Livestock preferentially graze the short annual plants, but within a few months after the end of the wet season they are forced to rely on the less palatable and less nutritious perennial grasses. Despite the inhospitable nature of these hot, shadeless plains with forage of low nutritive value for a good part of each year this country supports several times the intensity of stocking of any adjacent land types. It carries a substantial proportion of Australia's arid land flocks and herds. With continuous heavy stocking the perennial grasses are reduced or eliminated and the pasture becomes an annual grassland lacking the reserve forage provided by perennial grasses in the late dry season. Even when devoid of vegetation the flat topography and fine-textured soils are resistant to erosion.

Chenopod shrublands

These occur in the southern half of the zone where, on the average, summer and winter rain are about equal. The most common dominant plant species are *Atriplex vesicaria* (bladder saltbush) on cracking clay or texture-contrast soils and *Maireana sedifolia* and *M. pyramidata* (bluebushes) on calcareous soils. These perennial bushes are low (< 1 m) and spreading. Annual grasses and forbs grow in the interspaces in favourable periods.

These lands are mostly used for grazing by sheep which concentrate on the short annual vegetation when it is present, but rely on browsing the shrubs in other periods. Under heavy grazing the shrubs decrease and are replaced by shorter, less useful plants, particularly *Bassia* spp. The vegetation has been depleted over a considerable proportion of these lands, and much has suffered erosion of varying degrees.

Low woodlands

These occur on medium to fine-textured soils throughout most of the zone except the far north. Characteristically they are dominated by tall shrubs or low trees (2–10 m high), mostly various *Acacia* spp., of which *A. aneura* (mulga) is the most widespread and best known. The understorey varies greatly from place to place, even over short distances. In places it is dominated by perennial grasses such as *Eragrostis* spp. and *Monochather paradoxa*, in others by short annual (or short-lived perennial) grasses such as *Aristida* spp. and *Enneapogon* spp., and in the south by perennial or annual chenopodiaceous plants.

Low woodland country is used for grazing by sheep or cattle, both of which preferentially graze the short grasses and forbs when they are present. Most of the dominant tall shrubs and low trees are browsed by livestock in unfavourable periods. The effect of past grazing on the perennial shrubs varies in different regions. In the arid winter rainfall areas shrubs have decreased, in arid summer rainfall areas shrub populations appear little affected. In higher rainfall parts, particularly in the east of the zone, grazing has caused such an increase in shrubs, mostly unpalatable to livestock, as to render the land almost useless for grazing.

Spinifex grasslands

Vast areas of sandplain and dunefields are characterized by *Triodia* spp. or *Plectrachne* spp. (collectively called spinifex) which are large evergreen perennial tussock grasses with rigid, spiny, unpalatable foliage of low nutritive value.

Most spinifex areas are not used for livestock but small parts, with 'softer' species, are grazed at low intensities particularly in conjunction with other, better types of country.

Floodplains and alluvial fans

Floodplains and alluvial fans occur as small discontinuous areas adjacent to and at the end of rivers and creeks and at the foot of mountains and hills. Their total area is small compared with the previous four land types, but with their favourable water régime and commonly fertile soils they are very productive. Their soils and vegetation are very diverse, but they are preferentially grazed by livestock to the extent that they support grazing intensities possibly 100-fold greater than adjacent lands. It is not surprising that many are severely degraded and eroded.

Mountains and hills

Much of the relatively small areas of mountains and hills is too steep and rugged to be used for livestock, but the lower and gentler slopes are grazed at similar intensities to adjacent land types.

Exploration and settlement

The first Europeans to see any part of arid Australia were Dutch traders who happened to sail too far east on their journey to Indonesia. Their accounts emphasize the desolate nature of the western coast. The first British visitor was William Dampier in the late 17th century. His description of the land and its inhabitants was depressing.

In 1788 the first European settlement in Australia was established at Sydney, but it was three decades later before the first explorers from the settlement crossed into the arid lands. With the establishment of a colony at Adelaide in 1836 the tempo of exploration increased. New land was required for the colonies' increasing flocks and much of the exploration, both official and private, was inspired by pastoralists. The evaluations in this period were over-optimistic. First, it is likely that these early explorers, and those to follow in the next few decades, were only able to penetrate the inland areas during favourable years. Abortive expeditions were probably not recorded, thus biasing evaluations towards favourable years. Secondly, there were undoubted exaggerations; from reading some of the journals in the light of present-day knowledge, expressions such as 'passed through grass up to the horse's belly' should be interpreted as 'crossed a creek line today.'

Thirdly, the population and colonial administrators selectively interpreted the journals; they made much of the favourable parts, but neglected the descriptions of the poorer parts even when explorers suffered incredible hardships in traversing them. For example, despite the fact that Stuart had to make three attempts before successfully crossing Australia from Adelaide to the north coast and that he reported difficulties in finding water for his lightly equipped, mobile, horse-mounted expedition, the Governor of South Australia reported to the Colonial Office that cattle and sheep could easily be driven over the whole route (Duncan 1967).

By the early 1860s most of the grazeable land in arid Australia had been located and described by explorers and settlement began in earnest. Most of the whole vast area of grazeable land was occupied within the next three decades. The new settlers moved out from higher rainfall areas into areas with lower and less reliable rainfall, which perhaps partly explains their overestimation of the stock carrying capacity of the new lands and their underestimation of droughts.

Initially, crown land was occupied by the simple process of 'squatting' without legal sanction, but the situation was soon legalized by granting leasehold. The initial settlement and subsequent development of a region astride the New South Wales–Queensland border has been described well by Heathcote (1965); other regions differ only in degree. The crown lands were vested in the various Australian State Governments, who were anxious to avoid nomadism on the one hand, and the development of vast land monopolies, on the other. A compromise was developed; the governments leased the land to individuals and companies on a long term basis. The original leases were as short as eight years, but they are now much longer, mostly 30 to 50 years (Heathcote 1969). Thus the real owners of the arid grazing lands in Australia are the State Governments who administer them through their various Departments of Lands.

Within three decades the grazing industry in the earliest settled parts (western New South Wales) was in difficulty (Barnard 1969). The sheep population had increased rapidly to a high level as had the rabbit population. Undoubtedly the high grazing pressure had caused the land and vegetation resources to degenerate. With the onset of drought the sheep population in part of western New South Wales crashed dramatically from 13.6 to 3.6 million (Butlin 1962; Perry 1967*a*). Since that time stock numbers have increased, probably in response to additional stock watering points bringing more land within grazing range, but they have never reached the early high level. Stock populations in other arid regions have followed a similar pattern; an initial rapid increase to high levels, a sudden crash, and then a slight recovery (Perry 1968).

A broad general view of the present condition of Australia's rangelands (Newman & Condon 1969) reveals that the more sensitive land and vegetation types have degenerated to poor condition, the more robust types remain in good condition and intermediate types are in moderate condition. It is likely that the high stocking rates following shortly after settlement caused the major degeneration and that resource condition has changed little since then. If this is so, the heavy stocking for a few years caused a long-lasting deterioration in the resources; with consequent lower safe stocking capacity. Had the early stock populations not been so high, present resource condition and safe stocking capacity probably would be higher.

In evaluating Australia's arid grazing lands the explorers, administrators and pastoralists overestimated their potential. These overoptimistic attitudes have been, at least partly, responsible for resource degeneration. The main reason for overoptimistic evaluations is a set of perceptions some of which still persist today.

(i) The land and vegetation was perceived as being able to support more stock than it could.

(ii) The normal climate was perceived as being that of good years as expressed by 'We haven't had a normal rainfall for 25 years'.

(iii) The degeneration caused by heavy stocking was perceived as being less than actual and was commonly blamed on other factors, such as rabbits and droughts.

(iv) Land and vegetation were perceived as being more resilient and recovery more rapid than actual as expressed by 'It always comes back after a good rain'.

Present grazing industries

Nearly half of arid and semi-arid Australia (mainly the spinifex sandplains and dunefields) is unoccupied by European man and his livestock. Most of the remainder was settled for livestock grazing between 80 and 120 years ago, which is about the same time that western U.S.A. was opened up for the same purpose. In Australia the gold won from the goldfields of Victoria and New South Wales in the few decades after 1851 provided much of the development capital for extension of settlement into the pastoral areas.

The arid and semi-arid area supports about 5 million cattle, mostly in the north, and 30 million sheep, mostly in the south (between 20 and 30 % of Australia's total sheep and beef cattle population) (figure 1). The livestock population of arid Australia has remained relatively constant since 1910, but its proportion of the total Australian population has fallen as livestock populations have increased in the higher rainfall areas due to the development of sown and fertilized pastures (Barnard 1969). To some extent the demand for livestock caused by improved pastures in the higher rainfall areas has been supplied from the arid areas.

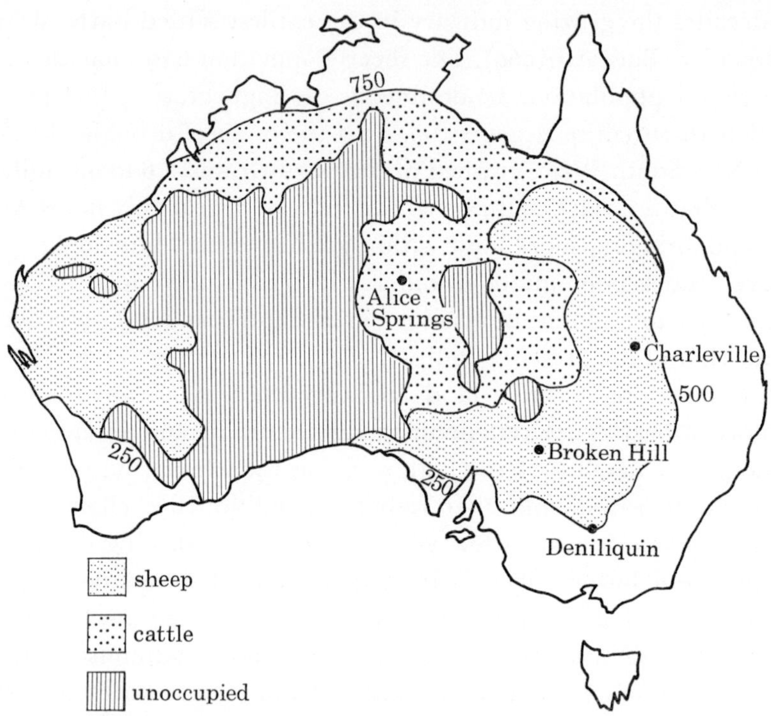

FIGURI 1. Map of Australia showing land use in the arid and semi-arid areas. The figures on the boundary of the arid and semi-arid area are mean annual rainfall in millimetres.

The livestock industries in arid Australia have always been oriented to export markets (Perry 1970*b*). In the past, annual variations in production from these areas had a very considerable influence on Australia's export income, but the upsurge of mineral exports has reduced the effect in recent years. Even today, with mineral exports accounting for about half of Australia's export income, the pastoral areas contribute about one quarter, and thus remain an important factor in the affluence of all Australians. Because of the export orientation, profitability is determined by international markets, and thus is affected by political and economic conditions in other countries. Profitability of Australian arid land grazing industries is affected by factors such as Britain joining the European Economic Community, the state of the Japanese economy, and the activities of the beef lobby in U.S.A.

By world standards, individual properties are very large. Sizes vary greatly, but 200000 ha is about average for cattle, and sheep properties are 25000–50000 ha in New South Wales and larger in other states. It is generally considered that herds of 2000 breeding cattle or flocks of over 5000 sheep are needed for economic viability. Profitability per unit area is very low (of the order of 1 to 3 cents (Aust.) per ha) and productivity per animal is low compared with higher rainfall areas but not with other arid lands.

Management is very extensive (as opposed to intensive); the relative absence of animal pests and diseases assists in this regard, as does the yearlong continuous grazing system. Apart from regular inspection to ensure that water supplies are adequate, the sheep and cattle are left to fend for themselves for most of the year, their distribution being controlled by fences and watering points. Cattle are mustered once or twice per year to brand calves and draft off beasts for market, bulls remain in the herd yearlong. Sheep are mustered for shearing and crutching and lamb marking, breeding is controlled by putting rams into the flocks for only a few months. This extensive management enables the industries to be very efficient in terms of labour – each man is able to manage about 1000 cattle or 5000 sheep, the use of fences instead of herders to control distribution being an important factor.

This efficiency in terms of labour can be illustrated further by considering total figures. In the 1966 census the total human population of pastoral Australia was 323000 (including 35000 aborigines). Of the total about 200000 live in mining and tourist centres, leaving only about 100000 concerned with livestock industries (this includes services to the industries as well as pastoralists themselves). As this 100000 earns about one quarter of Australia's export income, they are 40 to 50 times more productive (in terms of export income) than the average Australian.

In recent years rising costs of labour and the squeeze between costs and prices have caused a reduction in labour in the pastoral area. Migration to the coastal cities and towns has resulted in a steady decline in the rural population in the pastoral zone, and this decline is predicted to continue. Although the population is declining the efficiency per unit of labour is increasing.

The most important differences in the exploitation of Australia's arid and semi-arid lands in comparison with developing countries are socio-economic. The Australian grazier manages a highly capitalized, highly mechanized business reasonably well integrated with higher rainfall areas through a well organized communication, distribution and marketing infrastructure. Success or failure in Australian grazing industries depends on profits which is very different from the subsistence (or near-subsistence) industries of many other arid areas where failure means starvation for operator and family. Unlike his counterpart in most other countries, the Australian grazier enjoys a high status in the community. It is not long ago that the highest status in Australia was accorded to the owners of sheep stations; Australia 'rode on the sheep's back'. The situation has changed, but the Australian pastoralists remain a well-respected, well-educated, technologically advanced, and a relatively well organized, politically active group.

Scientific evaluation

Little applied research, with the objective of developing management strategies to maintain or improve the productivity of arid Australia, was done before the end of the Second World War. A considerable number of scientific papers had accumulated, but the literature was fragmentary, mostly reporting on short-term, isolated investigations in specific fields. This, and similar literature produced more recently, reveals that scientists, particularly the more academic of them, have different perceptions from graziers of the land and industries. Some of their perceptions which deserve questioning are:

(i) *Graziers are avaricious exploiters of the land and vegetation, mining the resources in chasing profits.* This is doubtful; certainly graziers have been the agents of resource degeneration, but more through their overoptimistic perceptions of the land and climate and an ignorance, shared by administrators and scientists, of good management principles.

(ii) *The resource degradation caused by past grazing is continuing at the same rate and will ultimately result in a vast dustbowl.* It is more likely that degeneration has been caused by one or more relatively short periods of over-use in the past and that the lands are now in a new equilibrium, albeit with many areas in poor condition and all areas subject to further degradation if subjected to future periods of misuse.

(iii) *Arid ecosystems are fragile.* An alternative suggestion is that many arid ecosystems are quite robust, the fact that they have degenerated being due to severe misuse rather than inherent fragility.

Within the last two decades applied research has increased and, for the first time, small research centres have been established at various locations within the arid and semi-arid zone. Scientists at these centres, and many others from institutions located in the higher rainfall areas, are doing research on biological and physical aspects of rangeland ecosystems (Christian & Perry 1969). In addition the Bureau of Agricultural Economics and economists in various universities have increased economic research on arid lands. Better definition of practical objectives for research has led to better integration, and investigations have begun on aspects such as resource inventory, resource condition and trend, safe stocking capacities and drought management strategies. The need for associating social, economic, and technological research has been recognized.

(*a*) *Resource inventory*

Shortly after the end of the Second World War, C.S.I.R.O. began a series of broad scale integrated land surveys in northern Australia. With the large areas to be covered speed was essential. A technique was devised of using interdisciplinary teams (geologist, geomorphologist, pedologist, plant ecologist), relying heavily on aerial photograph interpretation. These teams surveyed about 125000 km^2 each year mapping (generally at 1:1000000) and describing land systems defined as recurring patterns of topography, soils and vegetation (Christian & Stewart 1964). Within 15 years the nature and distribution of land types were described in much of the northern part of arid Australia (Christian *et al.* 1954; Mabbutt *et al.* 1963; Perry 1960; Perry *et al.* 1962; Perry *et al.* 1964; Speck *et al.* 1964; Stewart *et al.* 1970.) The broad scale was reasonably appropriate to the extensive nature of the pastoral industries (Perry 1967*b*).

More recently Queensland, Western Australia and New South Wales have adopted and modified the approach and are steadily completing an inventory of their arid grazing lands (Payne, Kubichi & Wilcox 1974; Wilcox & McKinnon 1974; Queensland Department of Primary Industry 1974). Some of these more recent surveys go further than inventory and include assessments of land condition and recommendations for administrative and legislative action.

(*b*) *Range condition and trend*

Since its inception in the late 1960s the C.S.I.R.O. Rangeland Research Unit (now part of the Division of Land Resources Management) has been concerned with establishing the concepts of range condition and trend in Australia and developing or adapting methods for assessing them. The rationale is that successful management of rangelands for long term productivity must be based on the condition and trend of the land resources rather than on short term profits or stock condition (Perry 1974, 1967*a*).

These concepts are new to Australia, but were developed in U.S.A. over 50 years ago and have been used there ever since. The reasons why they have been used for so long in U.S.A., but have only recently been introduced into Australia, are probably associated with differences in land tenure, responsibilities of Agencies, and population distribution. They were first developed in U.S.A. for Federal lands by Agencies charged with administering the lands in the public interest. Considerable centres of population near the lands and interested in uses other than grazing created pressure for maintenance of the lands in good condition. This urban public had a different perception of the effect of grazing on the lands than the ranchers. In Australia the lands are owned by State Governments, are administered by Agencies with charters not so specifically defining public interest, and are remote from any large and interested public. The reasons are social rather than technological.

The concepts of range condition and trend as important or essential rangeland management tools have already gained considerable acceptance by scientists in Australia and there are signs of interest among some of the administering agencies.

The common, but not the only, concept of range condition is analogous to health of an individual. Like health, condition involves many factors and is therefore difficult to assess absolutely. After some years of experimentation and experience Australian scientists are beginning to accept the American conclusion that the most appropriate single index to condition is change from original botanical composition. It is likely that future methods for assessing range condition in Australia will be based, as they are in U.S.A., on this single index.

Essentially range condition is the result of past management treatment and is therefore not a sensitive indicator of present or recent past management. The real indicator of recent management is range trend (current change in range condition). Range trend is more difficult to evaluate than range condition, but the development of suitable methods for assessing it, and their application in the management of Australia's arid grazing lands is probably the most important challenge to Australian range scientists in the near future.

(*c*) *Stock carrying capacity*

Ideally, in arid grazing lands, stock numbers should be flexible and managed according to observed range trend. From this point of view the idea of determining stock carrying capacities for various types of country has the inherent danger that the numbers so determined become a fixed right irrespective of climatic variations and of the effects on the resources. However, the practical situation is that administering agencies desire carrying capacity estimates for various purposes. The inherent danger is low if they are regarded as guidelines to be adjusted according to climatic and resource changes and not as long term absolute figures.

The New South Wales Soil Conservation Service devised a simple method for use in central Australia (Condon 1968; Condon, Newman & Cunningham 1969*a*, *b*, *c*, *d*). The principles of the method have since been applied to other regions. An inventory of land types (in Australia the land system surveys are used) and the location of stock watering points (to determine areas of land types within grazing range of water) are prerequisites for using the method.

Basically the method consists of:

(i) The best known land type under a particular mean annual rainfall and in good condition is chosen as a standard and allocated a stock carrying capacity, from experience.

(ii) For this standard land type, the factors which influence stock carrying capacity (topography, soil, hilliness and salt lakes, tree density) are all given a rating of 1.0.

(iii) For other land types a relative rating is estimated, from experience, for each of the factors. The product of these factor ratings provides a relative index for each land type in comparison with the standard.

(iv) Stock carrying capacities for each land type under standard rainfall are determined by multiplying the appropriate index by the carrying capacity of the standard.

(v) Stock carrying capacities for parts of land types under different rainfalls are adjusted by multiplying by a rainfall factor.

(vi) The condition of each occurrence of a land type (e.g. on particular properties or in particular paddocks) is allowed for by multiplying the carrying capacity by a condition rating factor.

The method provides a simple, practical way to determine stock carrying capacities. It uses only experience and the best estimates available, but is easily upgraded with increasing knowledge. Despite its simplicity and subjectivity I think the estimates obtained are probably as accurate as more complex methods requiring more sophisticated information.

In practice in Australia a stock carrying capacity in drought periods is also determined in a similar way but considering only drought reserve forage (mostly browse shrubs) and the factors influencing it. Using the 'normal' and 'drought' stock carrying capacities estimates of carrying capacities for the first, second, third and fourth years of a drought are recommended. Generally these are determined as follows:

(i) first year of drought – reduce stock by one quarter of the difference between 'normal' and 'drought' carrying capacities; (ii) second year of drought – reduce stocking by one-half the difference between 'normal' and 'drought' carrying capacities; (iii) third year of drought – reduce stocking to 'drought' carrying capacity; (iv) fourth year of drought – reduce stocking to one-half of 'drought' carrying capacity.

(*d*) *Droughts*

In most of arid Australia droughts are periods of forage shortage, water shortage only being a problem in areas where underground water is not available. Non-droughts are periods of adequate forage. It follows that management, and particularly stocking intensity, between droughts affect the duration and severity of drought periods.

In the Alice Springs district (central Australia) forage production depends on summer rains and non-drought years are those which follow summers during which the growing season is long enough to produce enough forage to last until the following summer.

The length of a growing season can be estimated from daily rainfall records with a simple soil water balance model (McAlpine 1970). By comparing estimates of growing season so calculated with 20 years of records of forage availability at five locations in central Australia I (unpublished data) was able to show that summers with a growing period of four weeks duration or more were followed by non-drought years and that summers with growing periods of less than four weeks duration were followed by drought years. Applying this criterion to the 92 years of rainfall records then available for Alice Springs, the probability of any year being a drought was estimated as 0.27. By using this figure and the Poisson distribution, droughts of one year duration can be expected in Alice Springs 14 times in 100 years, droughts of two years duration four times in 100 years, droughts of three years duration once in 100 years, droughts of four years duration once in 360 years, and droughts of five years duration once in 1250 years.

(*e*) *An economic computer model*

The Australian Bureau of Agricultural Economics (1974), in conjunction with officers of C.S.I.R.O., has developed and tested a computer model simulating the physical and financial functioning over time of beef production properties in the Alice Springs region. The model avoids the problem of the effects of different grazing intensities on the degradation of land resources by assuming that the 'normal' and 'drought' grazing capacities calculated by the method described above and the recommended drought destocking policy constitute conservative management and thus are compatible with long term resource stability.

The model is designed to investigate the effects on economic viability of such aspects as herd size, age of turnoff, beef price, availability of credit, borrowing rates, and levels of fixed and variable costs. It can be used to evaluate 'trade-offs' between variability in income and absolute size of income for particular properties and the probability of a property exceeding certain thresholds of income over a series of years with a normal sequence of drought and non-drought years. One result indicated by the model is that the minimum herd for long term financial viability is about 2500 breeding cows in the Alice Springs district.

Conclusions

For the last 100 years the grazeable part of arid and semi-arid Australia has been used for grazing sheep and cattle under a very extensively managed, market-oriented system, producing wool and meat mostly for export. Characteristically the industries have a very low productivity per unit area, a low productivity per animal, a high return to capital, and a very high production per unit labour.

The development and consolidation of the industries has been a trial and error procedure with overoptimistic perceptions of the productivity and resilience of the land resources. In the process resources have degenerated, some land types have been severely degraded, but others are little affected. The present rate of degradation is debatable. It is only in the last 20 years that integrated research with long term objectives has started.

Apart from the economic aspects of costs and market prices, the future viability of the grazing industries depends on maintaining the basic resources in a healthy state. Management practices based on the condition and trend of the land and vegetation resources need to be developed and applied. If safe management practices are developed and applied there are no technical reasons why the existing industries should not continue in the long term. Economic aspects may dictate some changes, and possibly some diversification of products.

Biologically and physically Australia's arid and semi-arid lands are comparable with those of other parts of the world, but there are tremendous social and economic differences. Technically, there is little problem in adapting successful management practices developed in Australia and U.S.A. to developing countries; the knowledge and experience are available. It is the large social and economic differences which appear to pose the major problems to their application.

References (Perry)

Barnard, A. 1969 Aspects of the economic history of the arid land pastoral industry. In *Arid lands of Australia* (eds R. O. Slatyer & R. A. Perry), pp. 209–228. Canberra: Australian National University Press.

Butlin, N. G. 1962 Distribution of the sheep population: preliminary statistical picture, 1860–1957. In *The simple fleece* (ed. A. Barnard), pp. 281–307. Melbourne University Press.

Bureau of Agricultural Economics 1974 A computer model simulating extensive beef cattle production systems. Occasional paper no. 21, 80 pp. Canberra: Bureau of Agricultural Economics.

Christian, C. S. *et al.* 1954 Survey of the Barkly region, Northern Territory and Queensland, 1947–48. *Land Res. Ser. C.S.I.R.O. Aust.* **3**, 1–182.

Christian, C. S. & Perry, R. A. 1969 Arid land studies in Australia. In *Arid lands in perspective* (eds W. G. McGinnies & B. J. Goldman), pp. 208–225. Tucson: The University of Arizona Press.

Christian, C. S. & Stewart, G. A. 1964 Methodology of integrated surveys. *Conference on Principles and Methods of Integrated Aerial Survey Studies of Natural Resources for Potential Development.* Paris: U.N.E.S.C.O.

Condon, R. W. 1968 Estimation of grazing capacity on arid grazing lands. In *Land evaluation* (ed. G. A. Stewart). Melbourne: Macmillan of Australia.

Condon, R. W., Newman, J. C. & Cunningham, G. M. 1969*a* Soil erosion and pasture degeneration in central Australia. I. Soil erosion and degeneration of pastures and topfeeds. *J. Soil Conserv. Service New South Wales* **25**, 47–92.

Condon, R. W., Newman, J. C. & Cunningham, G. M. 1969*b* Soil erosion and pasture degeneration in central Australia. II. Prevention and control of soil erosion and pasture degeneration. *J. Soil Conserv. Service New South Wales* **25**, 161–182.

Condon, R. W., Newman, J. C. & Cunningham, G. M. 1969*c* Soil erosion and pasture degeneration in central Australia. III. The assessment of grazing capacity. *J. Soil Conserv. Service New South Wales* **25**, 225–250.

Condon, R. W., Newman, J. C. & Cunningham, G. M. 1969*d* Soil erosion and pasture degeneration in central Australia. IV. *J. Soil Conserv. Service New South Wales* **25**, 295–321.

Duncan, R. 1967 *The Northern Territory pastoral industry*, 1863–1910. Melbourne University Press.

Heathcote, R. L. 1965 *Back of Bourke: a study of land appraisal and settlement in semi-arid Australia.* Melbourne University Press.

Heathcote, R. L. 1969 Land tenure systems; past and present. In *Arid lands of Australia* (eds R. O. Slatyer & R. A. Perry), pp. 185–208. Canberra: Australian National University Press.

Leigh, J. H. & Noble, J. C. 1969 Vegetation resources. In *Arid lands of Australia* (eds R. O. Slatyer & R. A. Perry), pp. 73–92. Canberra: Australian National University Press.

Mabbutt, J. A. *et al.* 1963 General report on lands of the Wiluna-Meekatharra area, Western Australia, 1958. *Land Res. Ser. C.S.I.R.O. Aust.* **7**, 1–215.

McAlpine, J. A. 1970 Estimating pasture growth periods and droughts from simple water balance models. *Proceedings of the XI International Grasslands Congress*, pp. 484–487. St Lucia: University of Queensland Press.

Meigs, P. 1953 World distribution of arid and semi-arid homoclimates. *Reviews of Research on Arid Zone Hydrology*, pp. 203–210. Paris: UNESCO.

Newman, J. C. & Condon, R. W. 1969 Land use and present condition. In *Arid lands of Australia* (eds R. O. Slatyer & R. A. Perry), pp. 105–132. Canberra: Australian National University Press.

Payne, A. L., Kubichi, A. & Wilcox, D. G. 1974 *Range condition guides for the West Kimberley Area, Western Australia*, pp. 1–141. Perth: Western Australian Department of Agriculture.

Perry, R. A. 1960 Pasture lands of the Northern Territory. *Land Res. Ser. C.S.I.R.O. Aust.* **5**, 1–55.

Perry, R. A. 1967*a* The need for rangelands research in Australia. *Proc. Ecol. Soc. Australia* **2**, 1–14.

Perry, R. A. 1967*b* Integrated surveys of pastoral areas. *Second International Seminar on Integrated Surveys of Natural Grazing Areas.* Publications of the ITC–UNESCO Centre for Integrated Surveys, Delft, Netherlands.

Perry, R. A. 1968 Australia's arid range lands. *Ann. arid zone* **7**, 243–249.

Perry, R. A. 1970*a* Arid shrublands and grasslands. In *Australian grasslands* (ed. R. M. Moore), pp. 246–259. Canberra: Australian University Press.

Perry, R. A. 1970*b* Productivity of arid Australia. In *Arid lands in transition* (ed. H. E. Dregne), pp. 303–316. Washington, D.C.: American Association for Advancement of Science.

Perry, R. A. 1974 The future of arid land vegetation under grazing. In *Studies of the Australian Arid Zone.* Vol. II. *Animal Production* (ed. A. D. Wilson). CSIRO Division of Land Resources Management, Australia.

Perry, R. A. *et al.* 1962 General report on lands of the Alice Springs area, Northern Territory, 1956–57. *Land Res. Ser. C.S.I.R.O. Aust.* **6**, 1–280.

Perry, R. A. *et al.* 1964 General report on lands of the Leichhardt–Gilbert area, Queensland, 1953–54. *Land Res. Ser. C.S.I.R.O. Aust.* **11**, 1–224.

Queensland Department of Primary Industry 1974 Western Arid Region. Land Use Study, part 1. Queensland Department of Primary Industry: *Brisbane Technical Bulletin*, no. 12, pp. 1–131, plus appendices i–x.

Speck, N. H. *et al.* 1964 General report on lands of the West Kimberley area, Western Australia. *Land Res. Ser. C.S.I.R.O. Aust.* **9**, 1–220.
Stewart, G. A. *et al.* 1970 Lands of the Ord–Victoria area, Western Australia and Northern Territory. *Land Res. Ser. C.S.I.R.O. Aust.* **28**, 7–61.
Wilcox, D. G. & McKinnon, E. 1974 *A report on the condition of the Gascoyne Catchment.* Department of Agriculture, Western Australia, Perth.

Discussion

S. Sandford (*Overseas Development Institute, 10–11 Percy Street, London W1*). Dr Perry has described how, in Australia, the beginnings of a probabilistic approach to estimation of stock carrying capacity had been made, with different estimates for 'normal' years and for successive years of drought. Unfortunately in most estimates of stock carrying capacity in developing countries a single figure is still being used. A move towards a more probabilistic approach is overdue, with estimates being made in the form of x hectares per livestock unit in y years in every 10, but $x \times z$ hectares in $(10-y)$ years in every 10.

Dr Perry claimed that enough scientific and technical knowledge existed already to improve productivity in the semi-arid areas, and that the remaining obstacles are socio-economic. But if one used Dr Perry's own figures for changes in livestock populations in the pastoral zones of Australia over the last 60 years, and assuming that figures for population are a reasonable proxy for figures for output, then the evidence appears to suggest that not even in Australia, where socio-economic problems are less serious than in developing countries, are increases in productivity being achieved; and this must cast doubt on whether adequate scientific and technical knowledge does really exist.

R. A. Perry. Climatic factors limit productivity of arid land grazing industries to relatively low values per unit area. In the developed countries technical knowledge has been used mainly in:

(i) establishing standards for safe resource management;
(ii) substituting capital for labour.

The available technical knowledge could be adapted easily for the same purposes in underdeveloped countries but social and economic factors are very different.

Sir Joseph Hutchinson, F.R.S. (*St John's College, Cambridge*). Mr Perry has told us that the only real insurance the Australian grazier has against drought is money in the bank. The subsistence herdsman in the semi-arid areas of the Third World seeks insurance by maintaining the maximum number of animals on the hoof. Though he may argue that banknotes do not produce calves and his stock do, nevertheless in the long term his extra stock are a liability rather than an asset. In times of drought they aggravate the situation against which they are supposed to insure. This is the real argument for bringing subsistence herdsmen at least to some extent into the market economy. The offtake of meat or of milk products from grazing lands can be used to provide a real insurance against drought, in that the returns from it are earned in sectors of the economy not directly dependent on the vicissitudes of the climate.

Phil. Trans. R. Soc. Lond. B. **278**, 507–524 (1977) [507]
Printed in Great Britain

India: underground water resources

BY J. K. JAIN
Chief Engineer (MI),
Department of Agriculture and Member (Development), Central Ground Water Board,
Ministry of Agriculture and Irrigation, New Delhi, India

Irrigation is important in India where one-third of the land surface is semi-arid and the rainfall is seasonal and erratic. The irrigated area in the country has almost doubled during the last 25 years and now stands at 43 million ha. Groundwater contributes to 40 % of all irrigation. Apart from providing irrigation, groundwater has been contributing to irrigated agriculture in many other ways and has, therefore, become a vital factor in the country's plans for agricultural development.

The total usable groundwater resources of the country are assessed at 350×10^9 m^3. The factors contributing to these resources and governing their occurrence and distribution in different parts of the country are described. At present about 35 % of the available resources are utilized.

Different types of groundwater structures, typical to India and suited to meet the irrigation needs of small farmers, are being constructed to tap the available resources in various parts of the country. Their design and construction features, determined to a large extent by the local geological formation, are explained.

The first large-scale venture in scientific planning and the development of groundwater was undertaken in 1934. A major break-through came in 1965 with the advent of high-yielding varieties of crops. Two important factors, namely the expansion of the rural electrification programme and the development of the cooperative credit structure, have helped in sustaining a high tempo of development since then. The present dimensions and other features of the development programme are indicated.

A systematic programme of groundwater investigations is a must for scientific development and management of the resources. It has been greatly intensified during the recent years and investigations include hydrogeological survey, geophysical studies, exploratory drilling, pump tests, and water balance studies including the construction of mathematical models. The approach and methodology adopted to keep down the cost of investigations is described.

No amount of groundwater investigation and planning will help in preventing infructuous expenditure, unless there is legislation to control and regulate groundwater development. The steps taken in this direction are mentioned.

1. INTRODUCTION

The water resources of a country constitute one of its most important economic assets. The pattern of water resource development for various beneficial uses, however, differs from country to country depending on its climatic and physiographic conditions and socio-economic development. In India, at present more than 90 % of the total water usage is accounted for by irrigation (including associated hydro-electric power generation). This is so because 70 % of India's population is dependent on agriculture and, rainfall being highly seasonal and erratic, successful agriculture is not possible without irrigation. Although the irrigated area has almost been doubled since the inception of planning about 25 years ago, still about 75 % of the gross cultivated area is without irrigation. Moreover, agricultural production in the country is still short of requirement. It is, therefore, likely that while the requirements and priorities for industrial and domestic use of water will undoubtedly increase over the course of years, irrigation may still continue to claim the maximum share of water resource development.

The contribution of groundwater to total irrigation is 40 %. The most conspicuous development of groundwater for irrigation has been in the semi-arid region of the country. With the extension of cheap electric power for pumping, the old belief that because of the lifting of water involved, the economics of groundwater schemes is less favourable than of surface water irrigation projects, is no longer true. The higher cost of operation is more than compensated by the lower order of initial investment on the schemes. The latter, in fact, becomes a positive economic advantage when there is constraint of financial resources.

The rôle of groundwater in agricultural development is more important than merely increasing the irrigation potential. Most of the old canal systems in India were designed to provide irrigation of the protective type as an insurance against famine and scarcity. These systems are not able to meet the irrigation demand of modern productive agriculture. Groundwater development in the command of these projects has greatly helped to make good the deficiency of the irrigation supply. On many canal systems, it has been found to be a powerful tool in controlling waterlogging and salinization and in substantially reducing the need for horizontal drainage. Also, the experience during recent occasions when the country passed through widespread and severe droughts, has demonstrated to the farmers that groundwater is a more dependable and handy source of irrigation under such periods of stress. Moreover, the means of tapping groundwater, i.e. wells and tubewells, are generally owned by the cultivators themselves. With the control of water in their hands, cultivators have the incentive to make the heavy investment required by productive agriculture. Groundwater may thus rightly be said to be providing the basic infrastructure on which the edifice of modern agriculture is being built in India.

2. Groundwater resources – quantum and the contributing factors

The groundwater resources are dynamic in nature; they grow with the expansion of irrigation.

The main factors that build up the groundwater resources are: (*a*) rainfall; (*b*) seepage from the canal system including storage reservoirs and the return flow from irrigation; and (*c*) inflow from the rivers. It is obvious that the contribution from factor (*b*) will increase progressively with the augmentation of irrigation. The contribution of factor (*c*) is also likely to increase due to the lowering of the water table along the rivers as further groundwater development takes place and as a result of measures for induced recharge from the rivers.

The estimated contribution from these factors as at present and as may be obtained ultimately on full development of irrigation (the ultimate potential for irrigation development in the country is assessed as 107 million ha as against the present level of 43 million ha) is given in table 1 (Jain 1974*b*).

The total estimate of present and ultimate groundwater resources thus comes to 650×10^9 m^3 (650 milliard cubic metres) and 850×10^9 m^3 respectively. All these groundwater resources, however, cannot be utilized because of physiographic limitations, inevitable loss as evapotranspiration in some high water table areas and in view of the obligation to maintain the committed base flow in the rivers. Out of the present estimated figure of 650×10^9 m^3 for the total groundwater resources, the withdrawal use is estimated at 120×10^9 m^3. Assuming that the net annual change in groundwater storage is negligible, it is estimated that the remaining 530×10^9 m^3 is presently accounted for to the extent of about 80×10^9 m^3 by the loss due to evaporation from the water table and to the extent of about 450×10^9 m^3 by the effluent discharge to rivers and streams (the latter is the contribution of groundwater resources to surface water). On the presumption that the effluent discharge would be maintained ultimately more or less at the same level as at present (that is 450×10^9 m^3) but the loss by evapotranspiration may be somewhat reduced as a result of lowering of the water table due to further groundwater development (say from 80×10^9 m^3 to 50×10^9 m^3), the estimate of the ultimately usable groundwater resources is placed at 350×10^9 m^3.

Table 1. Assessment of groundwater resources (10^9 m^3)

contributing factor	at present	ultimate on full development of irrigation
rainfall	500	500
seepage from canal systems including return flow from irrigation	110	250
influent recharge from rivers	40	100
total	650	850

In comparison, the total surface water resources of the country are assessed at about 1780×10^9 m^3. The utilizable component, however, is estimated as only 666×10^9 m^3, of which about 250×10^9 m^3 are presently used.

It may be stated that all the above estimates are tentative and will be refined as more elaborate data are collected.

3. Occurrence and distribution of groundwater resources

While rainfall is the major source of groundwater recharge, infiltration and storage of groundwater are governed largely by the geological formations. These two factors together give a fair general indication of the occurrence and distribution of groundwater resources in different parts of the country.

(*a*) *Rainfall*

The average rainfall in India is about 120 cm, which is a little more than the global mean of 99 cm. However, it is highly seasonal in character and widely varying in distribution over the country. About 75 % of the total rainfall is received in the four-month period, i.e. June to September. Maximum groundwater recharge takes place during this period, and the groundwater contours are usually at their highest level immediately after this period.

The rainfall map of India (figure 1) shows that an average rainfall of the order of 100 cm and more occurs in about 50 % of the total land area of the country covering the Eastern Region, the Himalayan Region and the coastal areas. An appreciable portion of this region in the northeast and along the western coast receives rainfall of more than 200 cm. This region accounts for about 55–60 % of the total usable groundwater resources of the country.

On the other hand, the bulk of groundwater development that has taken place and made a significant contribution to stepping up wheat production in the country has been in the region receiving less than 100 cm of rainfall. A major portion, comprising about 60 %, of this region, towards northwest and covering the rain shadow area formed between the mountains, called the Ghats, along both the coasts in peninsular India, receives a rainfall between 40 and 75 cm and may be categorized as semi-arid. Most parts of this region, however, happen to be under heavy canal irrigation from snow-fed and other rivers and have good groundwater potential in spite of low rainfall.

The fringe area in the extreme northwest is an arid region receiving less than 40 cm of rainfall. Groundwater availability is by and large chequered in this region and the groundwater encountered is frequently of poor chemical quality. Saline and fresh water zones are

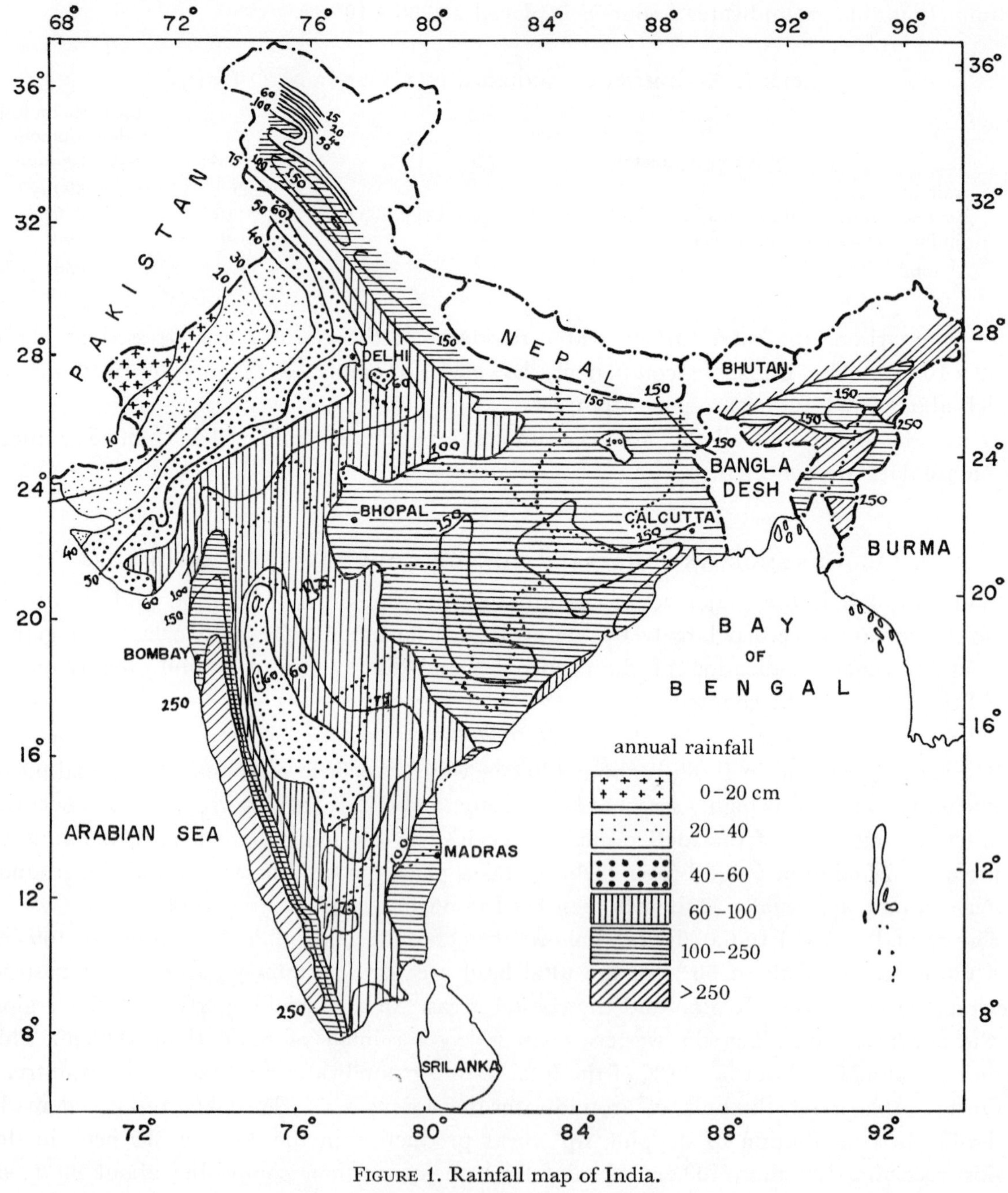

FIGURE 1. Rainfall map of India.

intermixed in a complex manner and this becomes a serious problem in the utilization of the available groundwater resources. In a part of this region, there has been located a sandstone formation of Jurassic age which holds a reserve of good quality groundwater (about 28×10^9 m^3). The water is estimated to be 7000 years old and has no current recharge.

(*b*) *Geological formations*

Groundwater availability is largely governed by the state of cementation and compaction of the formation. From this point of view, the geological formations encountered in the country may be broadly divided into three categories (Deshmukh 1975), namely, the unconsolidated; the semi-consolidated; and the consolidated (figure 2).

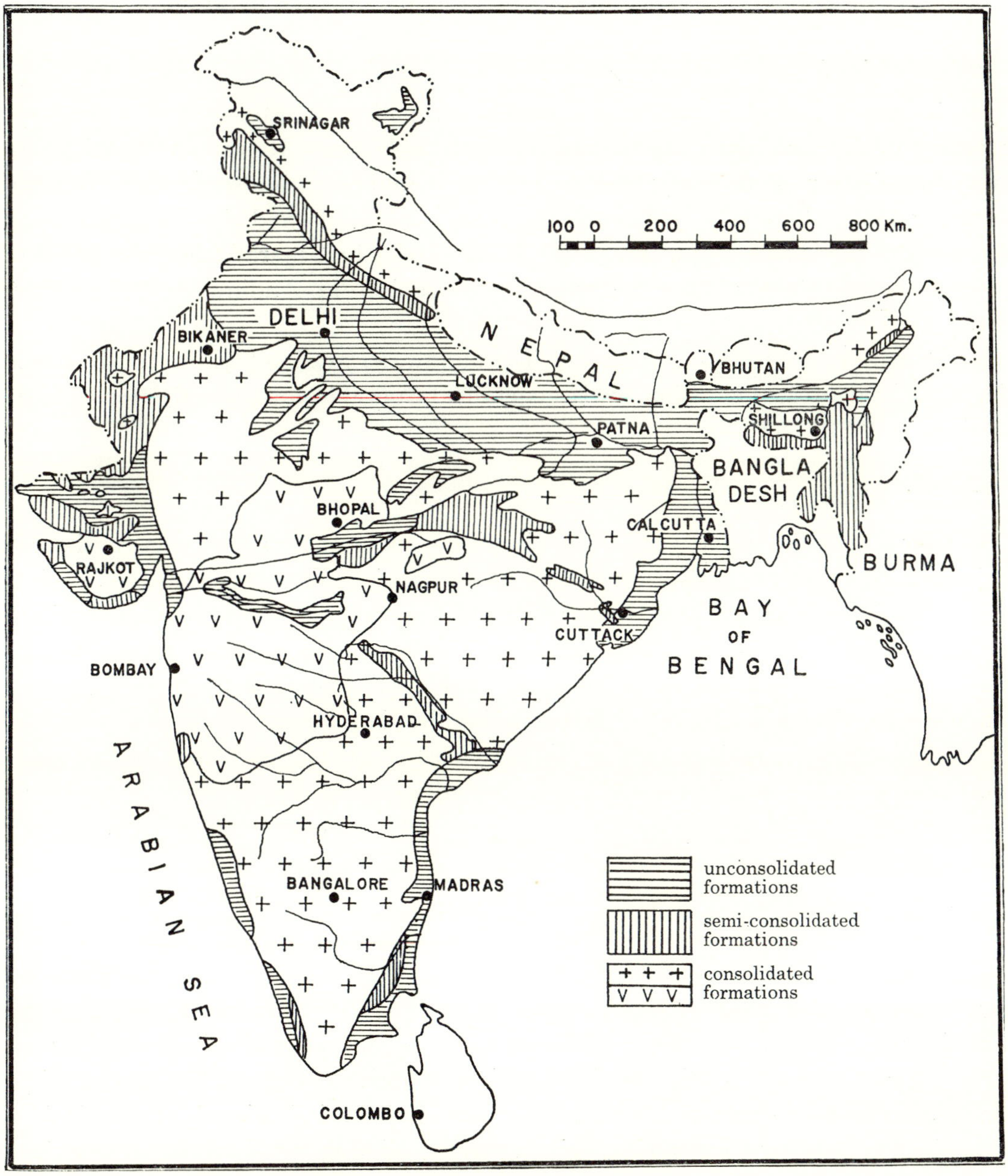

FIGURE 2. Hydrogeological map of India.

(i) *Unconsolidated*

These include: the alluvial plains built up by the Himalayan rivers, namely, the Indus, the Ganga, the Brahamputra and their tributaries; the inland valleys of the rivers Narmada, Tapi and Purna in central India; the coastal alluvia – of east and west coasts and plains of Gujarat; and the aeolian sand-tracts of western Rajasthan. These formations are richest in groundwater except in parts of the northwestern arid and semi-arid zones not served by surface irrigation canals. They comprise about one-third of the total land area in the country (which is 3251200 km^2), but account for about 50–60 % of the total usable groundwater resources. The lithology includes zones of sands, gravels, pebbles, etc. which not only store large quantities of groundwater but are also favourable from the point of view of groundwater extraction.

(ii) *Semi-consolidated*

These include the semi-consolidated sandstone formations of Mesozoic and Tertiary age, interspersed in different parts of the country, such as Western Rajasthan (Lathi series and Palana and Nagaur sandstones); east coast (the Cuddalore sandstones and Rajamundri sandstones); the west coast (the Warkalli formations in Kerala and the Himmatnagar sandstones and the sandstones of the Bhuj and Katrol series in Gujarat); and the rocks of the Gondwana systems (which have sandstones and shales) in Andhra Pradesh, Bihar and Orissa.

These formations are next in importance in groundwater availability, but hardly cover 5 % of the total land area. The lithology is generally favourable for groundwater storage and extraction.

(iii) *Consolidated formations*

The major part of the country including almost the entire Indian peninsula consists of hard rock formations. These comprise: crystalline rocks, granites, gneises, schists, etc. of Archean age; basaltic formations of Upper Cretaceous to Oligocene age; and compacts sedimentary formations – the Cuddapah, the Dehlis and the Vindhyan system which belong to the Precambrian age. Nearly two-thirds of the total land area is covered by these formations, of which about one-fourth is accounted for by the basaltic formations.

The rocks in these formations have no primary pore spaces and only hold limited quantities of groundwater in the weathered and the fractured zones. However, sizeable quantities are available at some locations in vesicular lava flows, intertrappean beds and cavernous limestones.

4. Groundwater structures – salient design and construction features

Diverse geological formations require different types of structures for tapping groundwater to meet irrigation needs. The choice of the structure in an area is also influenced by the size of farm holdings and the relative preference given to private, cooperative and public wells. Broadly, groundwater structures commonly constructed for irrigation in India (Jain 1973) may be divided into three categories: (*a*) dug wells; (*b*) dug-cum-bore wells; and (*c*) tube-wells.

(*a*) *Dug wells*

These comprise open surface wells of varying dimensions dug or sunk from the ground surface into the water bearing stratum. Usually two types of wells are constructed: (i) masonry (lined) wells; (ii) unlined wells in hard rock.

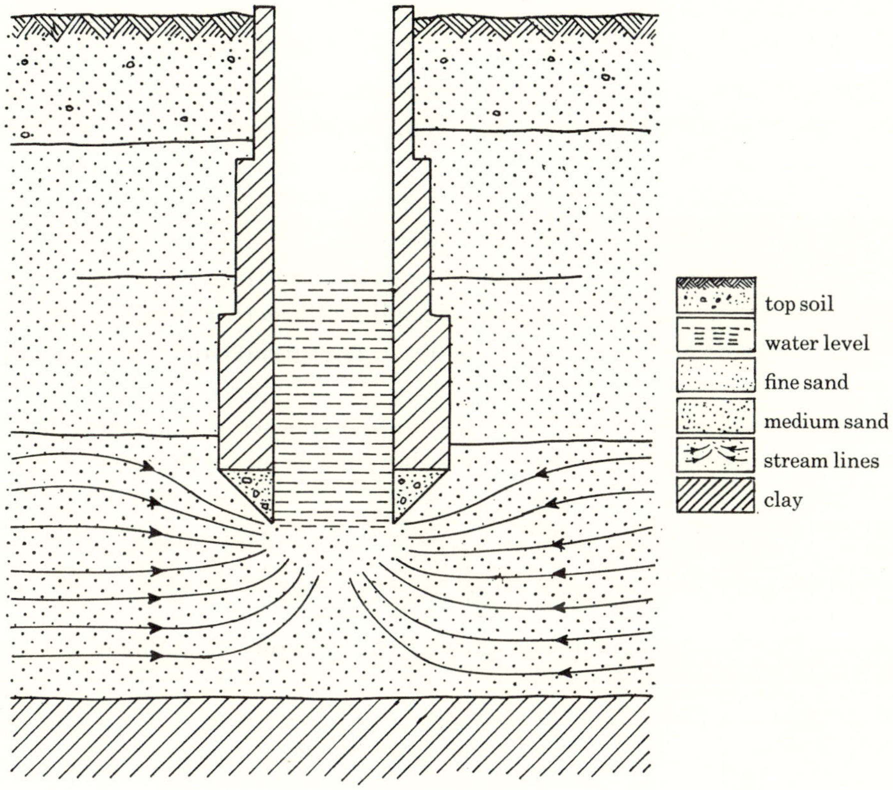

FIGURE 3. Masonry well in alluvium.

A typical masonry well usually constructed in the alluvial or semi-consolidated formations (figure 3) has a masonry steining wall sunk in subsoil by applying a static weight of sand bags and simultaneously scooping out earth from inside. In locations where it is not possible to sink masonry steining by ordinary methods due to the presence of hard clay or boulders, it is built in a dug pit from the bottom upwards. The diameter of the well ranges between 2 and 6 m, and the depth of the well between 8 and 15 m.

Research work is in hand in the country (Jain 1971) to improve the design of these wells in two respects. The draw-down and therefore the yield of these wells has to be restricted to prevent sand movement from the bottom so that the masonry steining does not collapse due to settlement. Attempts are, therefore, being made to design suitable inverted filters that may be provided at the foundation of the wells to control sand movement. Secondly, the flow in these wells is through the bottom only. This results in the congestion of stream lines and considerable loss of head. The wells are, therefore, hydraulically inefficient. Efforts are being made to evolve suitable cheap designs for pervious steining which should improve the efficiency and water output of these wells.

In alluvial formations with a shallow water table, both dug wells and shallow tubewells (described later) are feasible and the latter are generally cheaper in terms of cost of per unit of water produced. Dug wells, however, have the advantage that they permit a phased development to the poor farmer. In the first instance he goes in for the well only and then after earning some money he gets it bored and subsequently gets it fitted with a pumpset – which arrangement practically becomes a tubewell with the well steining serving as a pump house.

A typical dug well constructed in the hard rock formations is usually an open excavated pit through the top soil and the weathered rock. The top portion through the soil mantle is usually lined with masonry or dry stones. Pneumatic rock blasting equipment using jack-hammers and explosives is employed for excavation of the well through the hard rock. The well taps water from the sides as well as from the bottom and is generally constructed in a large size (6 m × 6 m or more) to provide storage.

The diameter and the depth of these wells is often decided arbitrarily. Case studies are in hand to assess the relative advantage of having a larger diameter (which increases the percolation area as well as the storage capacity) versus greater depth (which increases operational draw-down and therefore the yield and also the storage).

Persian wheels, rope-and-bucket lifts and sometimes counter-poised lifts are used to lift water from dug wells. The present trend, however, is in favour of using power pumps, preferably electrical but otherwise diesel if electric power is not available. The water output of these wells is of the order of 50–100 m^3/day and their cost ranges between \$250 and \$1000.

(*b*) *Dug-cum-bore wells*

Dug wells are frequently bored through the bottom to augment their output. These are referred to as dug-cum-bore wells.

In sedimentary formations, boring essentially consists of drilling a small size bore of diameter usually ranging between 7.5 cm and 15 cm, through the bottom of the well and extending the bore down to a layer of good water bearing medium with a view to tapping this aquifer. The boring is usually done by the percussion method, using hand boring sets or percussion rigs (Jain 1962). Bores are of two types: either cavity or strainer.

The dug-cum-bore well of the cavity type (figure 4) taps a water bearing medium, which is immediately below a hard impermeable layer, through a cavity formed in the medium as a result of gradual over-pumping. The bore is fitted with a plain pipe, extending down to the cavity. This type of boring is economical but is not feasible at all locations. An essential requirement is that the impermeable layer overlying the water bearing stratum should be sufficiently thick and strong.

In the dug-cum-bore well of the strainer type the water is tapped through a screen provided opposite the vertical depth of the water bearing medium. The rest of the bore is lined with unperforated pipe.

In hard rock formations again, bores are of two types: (i) bore tapping an embedded layer of water bearing medium; and (ii) bore tapping joints, fissures, fractures, etc.

Figure 5 shows the first type. The size of the bore is usually 7.5 cm or more. It is normally drilled by percussion or calyx rigs. Recently in some cases down-the-hole rigs have also been used for faster drilling.

In the second type the practice is to drill a number of slim holes, 4–5 cm in diameter

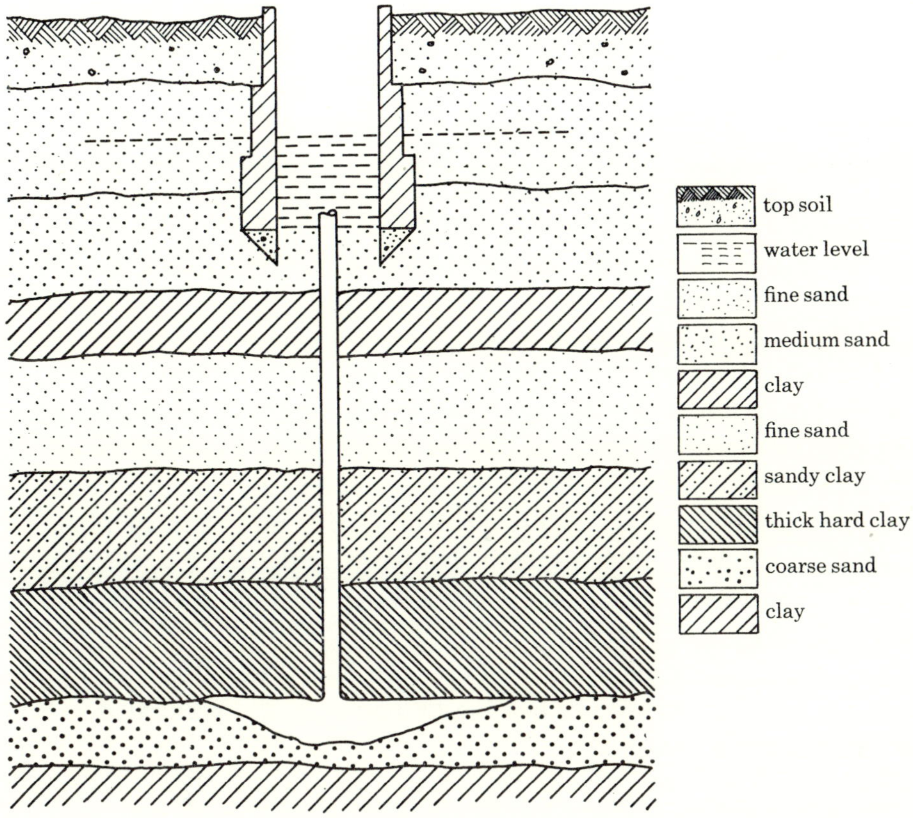

Figure 4. Dug-cum-bore well in alluvium (cavity type).

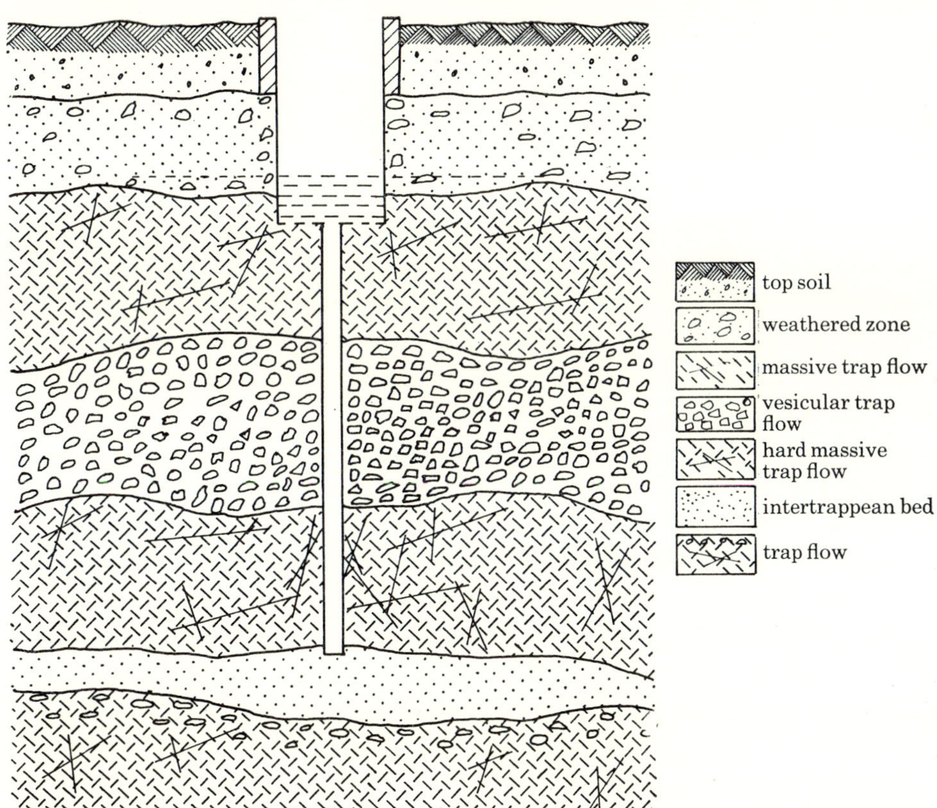

Figure 5. Dug-cum-bore well in hard rock (tapping intertrappean bed).

instead of one big central hole, in various directions perpendicular to the inclination of the water bearing fracture planes by using rock drills with extension equipment.

The water output of the dug-cum-bore well is considerably higher than that of a simple dug well. It is usually much more economical to operate these wells by pumpsets than by man or animal operated water lifts. The depth of the bore seldom exceeds 50 m and the extra cost involved in boring usually ranges between $100 and $600.

(*c*) *Tubewells*

A tubewell essentially consists of a borehole drilled into ground with the purpose of tapping groundwater from the pervious zone. Tubewells constructed in the country may be broadly divided into three categories: (i) shallow tubewells; (ii) deep tubewells; and (iii) bore wells. A tubewell in sedimentary formations not exceeding 60–70 m in depth is called a shallow tubewell. Shallow tubewells are of two types: cavity tubewells; and strainer tubewells.

In a cavity type shallow tubewell the water is directly pumped from a cavity formed at the bottom of the bore, which is provided only with unperforated pipe. Cavity tubewells are cheaper but a durable cavity is feasible only where there is a strong roof of hard impermeable stratum, immediately overlying a good water bearing medium, and the water bearing medium is comparatively thin.

In a strainer type shallow tubewell the bore hole is installed with a pipe assembly including plain pipes opposite non-water bearing formations and a perforated screen opposite the water bearing medium or media.

Shallow tubewells are usually drilled by percussion methods using hand boring sets and sometimes percussion rigs. They are generally operated by the centrifugal pumps (volute type). They normally operate for 6–8 h per day during the irrigation season and giving a yield of the order of 100–300 m^3/day – roughly 2–3 times that of dug wells. Their cost ranges between $500 and $1500.

They are usually privately owned by the individual farmers. Their success and popularity depends on how cheap they are. Research and development efforts are therefore being directed to evolve cheap strainers for these tubewells. Perforated PVC pipes and pre-gravel pack filters are some of the promising designs that are presently under trial. Meanwhile, coir strainers formed by winding coir string over an iron frame have become very popular, because of their low cost, and have replaced, to a considerable extent, the brass strainers that were formerly in use. In the shallow water table areas, even the use of the iron frame in the coir strainer has been replaced by the bamboo frame. Also, attempts have been made to replace the steel pipe casing with pipes constructed by wrapping bitumenized gunny bags over the bamboo frame. These are called bamboo tubewells and are very cheap.

The deep tubewells constructed in India usually extend to depths of 100 m and more, and are designed to give a discharge of 100–200 m^3/h. These are usually drilled by direct rotary or reverse circulation rigs except in the bouldery formations where percussion or rotary-cum-percussion rigs have to be used. They operate round the clock during the irrigation season. Their total annual water output is roughly 15 times that of an average shallow tubewell.

Their cost is as high as $10000–16000. They are, therefore, beyond the means of individual farmers and are usually constructed as public works, owned and operated by public corporations or Government departments. Water rates are charged for the service rendered to the farmers.

Usually deep tubewells in the country are constructed as artificial gravel packed wells (figure 6). There are situations where straight or naturally developed wells (strainer tubewells) will be more economical and efficient but still the practice is to go in for gravel packed wells. The important reason for this is that suitable screens with fine slots are not available in the country. Slotted steel tubes are mainly used as screens and it has hitherto not been possible to cut satisfactory slots of less than 1.6 mm size on these tubes.

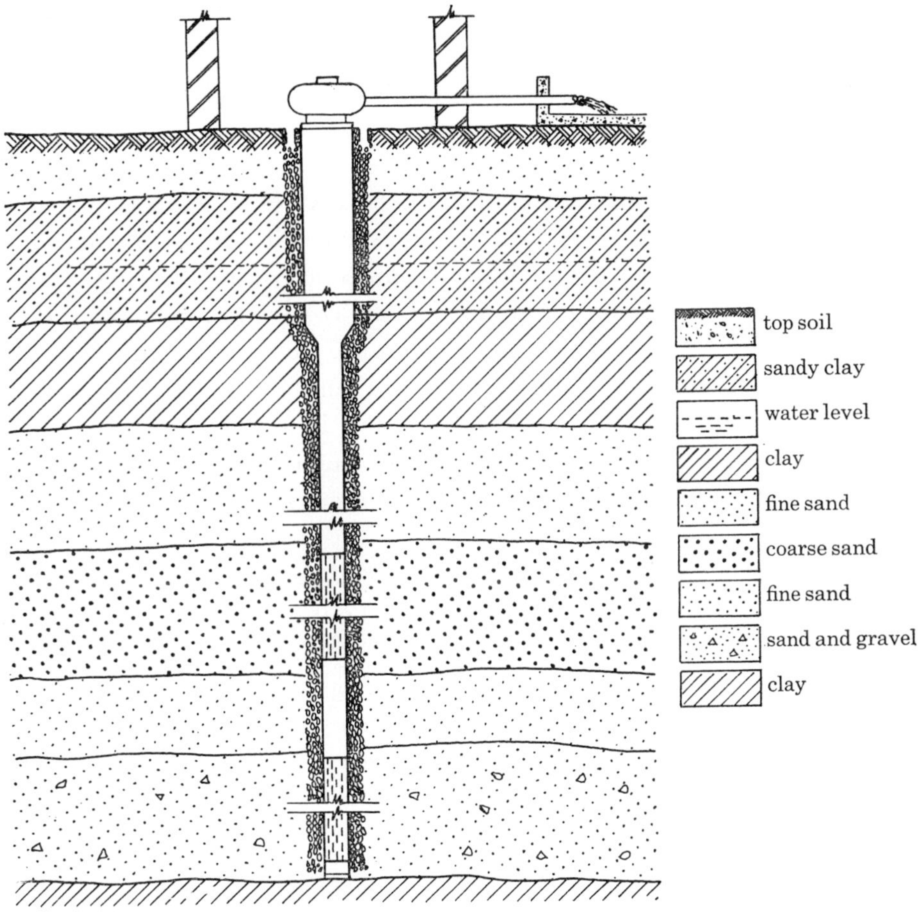

FIGURE 6. Deep tubewell.

Some far-reaching improvements (Jain 1973) have recently been effected in the design of these tubewells. These include:

(i) Designing the gravel pack and the slot size in relation to the aquifer material – definite criteria, based on research, have been evolved for this purpose. Further research work is in hand.

(ii) Designing the well for a high operational draw-down (the former practice was to limit the draw-down to about 6 m) so that the same well is able to deliver much more discharge at practically the same cost.

(iii) Increasing the open area of the screens from the previous 5–6 % to 15–25 %, as it is found that this improves the efficiency of the screens without endangering their structural stability.

(iv) Increasing the diameter of the screen assembly from 15–20 cm for wells discharging between 100 and 200 m^3/h in order to reduce the head loss.

(v) Reducing the area commanded by these tubewells so that they are able to provide intensive and assured irrigation required by the improved crop varieties now used.

The main criterion for the tubewell design still continues to be that the entrance velocity through the screens should not exceed 3 cm/s to control sand movement, encrustation and corrosion. Research work is in hand to evolve more refined criteria in this respect. Efforts are also being made to develop suitable alternative materials for the screens, particularly for use in formations where quality of water is a problem.

Tubewells in the hard rock areas are called bore wells because in such wells the borehole is stable without a lining in the bottom portion and a tube is put in only in the upper weathered zone.

These wells usually depend on the joints, fissures, fractures, etc. for their water supply. Even with a heavy draw-down of 20–30 m, such wells are usually not able to yield more than 5–10 m^3/h, except when they tap some embedded water bearing strata. Their operation is possible only with submersible pumps. Their cost ranges between $1200 and $1600 and is comparable to that of an average dug-cum-bore well fitted with a pumpset. Their advantages are that they do not occupy any agricultural land and can be completed in a very little time by hammer drills. Due to their meagre yield, however, they have not become popular for irrigation use in comparison to dug wells or dug-cum-bore wells, although they have established their merit for meeting drinking and domestic water requirements.

5. Programme for groundwater development – dimensions and policies

The use of groundwater in India through dug wells and indigenous water lifts dates back to ancient times. The first large scale venture in scientific planning and development of groundwater was undertaken in 1934 when a project for construction of 1500 deep tubewells was initiated by Sir William Stampe, the then Chief Engineer, in one of the important States of India, Uttar Pradesh. All the available knowledge on groundwater at that time was pooled and fully used in planning. The success of this project led to the extension of similar activity to other promising areas of the country. A significant breakthrough in groundwater development, however, came around 1965 with the appearance of high-yielding varieties on the agricultural scene and the realization that groundwater development is essential to provide the controlled and assured supply of irrigation needed to reap the full benefit of these varieties. The progress of development, in terms of important parameters, up to 1965 and afterwards, is set out in table 2.

Two major factors, namely, the expansion of the rural electrification programme and the development of the cooperative structure for financing privately owned irrigation works have played an important rôle in sustaining the high tempo of progress after 1965. During the fourth Five-Year Plan electric connections were provided to 270000 pumpsets every year. Progress slowed down during the last two years due to shortage of power and the limitations of main transmission and distribution lines. This, however, was a temporary phase and the programme has already started to regain its original pace.

The present annual expenditure on the programme of groundwater development is of the order of $400 million. About 60 % of the total amount is now mobilized through cooperative sources including land development banks and cooperative banks, which extend long-term credit (period 7–15 years) and medium-term credit (period up to 5 years) respectively. The remaining funds are partly provided from Government sources and partly arranged by the cultivators themselves. Financial support of the cooperative credit structure has indeed been a big factor in the expansion of the programme during the recent years.

TABLE 2. PROGRESS OF GROUNDWATER DEVELOPMENT (millions)

	1950	1965	1975
dug wells	3.50	5.00	7.25
tubewells	negligible	0.10	1.00
pumpsets	0.08	0.90	4.75
area benefited (10^6 h)	6.50	9.50	17.00

The annual programme presently embraces construction of about 300000 new groundwater structures. In addition, about 100000 existing dug wells are bored or deepened to augment their water output and fitted with pumpsets (in place of existing man or animal operated water lifting appliances) every year. In terms of volume of water, utilization of groundwater resources was of the order of 120×10^9 m^3 at the end of 1973–4, which marked the last year of the fourth Five-Year Plan. Utilization is presently increasing at a rate of about 5×10^9 m^3 per annum.

Although the programme of development has now been picking up momentum in all areas, the rate of development hitherto has been much faster in the areas receiving rainfall of less than 100 cm. In fact, a stage has been reached in many parts where much more sophisticated planning and development of groundwater resources in close integration with surface water development is necessary to avoid situations of overdraft, saline contamination, reduction of committed flows, etc.

About 85 % of the total groundwater development is taking place through dug wells, dug-cum-bore wells and shallow tubewells, which are cheap enough to be afforded by farmers on an individual basis and have enough water to meet the irrigation requirements of small farm holdings – the average size of holding in India is about 2.3 hectares. If any farmer owning a well has some surplus water, he sells it to his neighbours. Deep tubewells are much more expensive and are generally beyond the financial means of individual farmers. These are, therefore, generally owned and operated as public works. The policy adopted is that public tubewells are usually constructed in areas where the water level is deep or aquifers are deep seated or otherwise difficult to tap by shallow tubewells or dug wells. In areas where both deep tubewells and shallow structures are possible, deep tubewells are constructed as public works in such pockets as have a predominance of small farmers who cannot afford structures of their own.

The management of public tubewells had suffered from some administrative problems. These have been identified and serious efforts are being made to remove them. In several states the construction and management of public tubewells has been transferred to public corporations who are expected to give single-minded attention to these problems. These corporations are also able to mobilize institutional credit, in addition to the budgetary allocations by the Government, for undertaking the construction of additional wells.

For private works, in addition to the cooperative loan, grants up to one-third from

Government sources are also made available to small farmers having holdings less than 2 ha and up to one-half in cases where the small farmers combine together in a group to own and operate a well. Custom service in boring and drilling of wells as well as technical guidance in design and construction of wells is provided to the farmers by the State ground water organizations. There are also many small private agencies available to do the construction work and the farmer is free to choose between the departmental or a private agency.

Altogether the development of new irrigation potential, the problems of irrigation water management, and better utilization of existing irrigation potential are now receiving the serious attention that they deserve. A special programme called the Command Area Development Programme has been initiated for this purpose during the current Plan. This programme, however, will mainly cover public irrigation systems. As far as private wells are concerned, the standard of land levelling and shaping, constructing and maintaining farm distribution channels and crop planning, is already much better than on public irrigation systems. Further efforts to improve standards are being made by the lending agencies and the Extension Service.

6. Groundwater investigations – approach and methodology

That systematic groundwater investigations are essential for scientific development and management of groundwater resources hardly requires any elaboration. In fact, modern agriculture emphasizes the need for integrated development of ground and surface water resources (Jain 1974*a*) and this further increases the importance of scientific groundwater investigations and planning. The need is particularly urgent in areas where a critical stage has been reached and the hazards of over-development are in sight.

A systematic programme of groundwater investigations started in the country in 1954, with the launching of an all-India Exploration Project under the Technical Cooperation Programme of U.S.A. This project envisaged geophysical investigations to supplement the available geological data in regard to the structural disposition of the water bearing strata; and test drilling, including electrical logs of the bore holes and pump tests, to study the hydrogeological parameters in selected alluvial and sedimentary areas considered to be geologically favourable for the large-scale production of groundwater. This led to the setting up of the Exploratory Tubewells Organization (E.T.O.) under the Ministry of Agriculture and the creation of an exclusive wing in the Geological Survey of India (G.S.I.) for groundwater investigations. The two organizations did spade work in opening out new and promising areas for large-scale groundwater development. However, when the programme of groundwater development gained momentum around 1965, it was felt that there were deficiencies in groundwater investigations and planning (Jain 1975). It was noted that there was not a sufficient tie-up between exploration and the development programme. Exploration was mainly directed towards planning deep tubewells and was, therefore, largely confined to alluvial and sedimentary areas. Development on the other hand was taking place in almost all parts of the country and about 85% of total development was through dug wells, dug-cum-bore wells and shallow tubewells. Secondly, the main focus of the investigations had been on exploration and prospecting and the quantitative aspects of groundwater availability had lacked attention. Thirdly, while under the Indian Constitution, responsibility for the development of groundwater devolves on the States, they had no organization of their own for investigations and the planning of projects. Both the E.T.O. and the Ground Water Wing of the G.S.I. were located at the central level. It was further realized that the needs of the development programme were urgent and it would not be possible to wait for the results of very detailed and comprehensive groundwater investigations, which are not only expensive but also time consuming.

The programme of investigations has been intensified in the recent years and several corrective measures have been undertaken. It has been decided that no scheme or project will be sponsored by any financing agency unless it has been technically evaluated. In the absence of elaborate information, the project, unless it has some complicated implications, will be appraised on the basis of empirical norms and short-cut procedures. A guideline has been formulated for this purpose. The norms and procedures will be gradually refined as more data become available. To facilitate the use of data in formulating and appraising schemes, it has been agreed that compilation of data and information will conform to the district boundaries, which are the administrative units for planning. The two organizations at the central level, namely, the E.T.O. and the Ground Water Wing in the G.S.I., have been merged to form a unified organization, called the Central Ground Water Board, in order to bring more cohesion into the activities. Since the basic responsibility for development of groundwater resources devolves on the State Governments under the Indian Constitution, organizations have been set up in nearly all the states for planning schemes, collecting data that is relevant to planning from different sources including the on-going production programme, and undertaking short-term investigations to fill up gaps in the information.

A blue-print has been prepared for long-term investigations in the country to serve the increasing needs of the development programme at minimum cost. This visualizes extension of the basic programme of hydrogeological investigations to all parts of the country, including the areas under consolidated formations which had earlier been left out. These investigations include surface geological reconnaissance, photo-geological studies, geophysical investigations, well canvassing, water level measurements, and chemical testing of water. About half the total area of the country is expected to be covered by these studies by the end of the current Five-Year Plan. The programme will be of a continuous nature and the area once covered will be reappraised at regular intervals for up-dating the information. Expensive activities including exploratory drilling, geophysical logging and pump tests are to be undertaken in limited areas where these are essential to supplement other information. The need for water balance studies and special hydrological studies to provide quantitative assessments is fully recognized, but since such studies are expensive and time consuming, they are to be undertaken in limited representative basins so that the results achieved can be applied to larger areas. Nine projects for such studies are presently in hand. The scope for applying more refined techniques like remote sensing, radioactive tracer studies, mathematical models, etc. is being assessed under these projects.

The maximum emphasis under the strategy of investigation is on setting up grids of observation wells for continuous monitoring of the water levels and changes in the quality of water. These studies are expected to give most reliable information at minimum cost on the status of the water balance and the chemical régime of groundwater systems in a region at any particular time.

This strategy of investigation, coupled with research and development efforts in the engineering design of groundwater structures, has brought about considerable improvement. There is, however, full awareness that as groundwater development reaches a more critical stage (in fact it has already started doing so in some limited areas), much more sophisticated investigations and planning will be necessary. With the foundation already laid, it is hoped that the capability for the same will also be built up over time.

It is obvious that no amount of groundwater investigation and planning will help in preventing overdraft, saline infestation, reduction in committed base flow in the rivers and other hazards, unless there is legislation to control and regulate groundwater development. Realizing the importance of such a legislation, an important step has been taken in this direction, in that a model bill has been prepared and circulated to the states. The states are being urged to process it through their Legislatures at an early date. Fortunately, under the Indian Easement Act 1882, all groundwater vests in the government and no right to groundwater can be acquired by prescription. It is, therefore, felt that the proposed groundwater legislation should not pose any serious legal or constitutional problem.

References (Jain)

Deshmukh, D. S. 1975 Status of ground water studies, development and utilisation in India. Indo-German Workshop, N.G.R.I. Hyderabad (1–22).

Jain, J. K. 1962 *Handbook on boring and deepening of wells*, pp. 1–116. Government of India, Ministry of Food and Agriculture, New Delhi.

Jain, J. K. 1971 The need and scope for engineering research on design aspects of tubewells. *Annual Workshop of Indian Council for Agricultural Research, New Delhi, India*, pp. 1–28.

Jain, J. K. 1973 Ground water structures, pp. 1–58. Lecture delivered at the University of Roorkee, India.

Jain, J. K. 1974*a* Integrated development of water resources for irrigation. Lecture delivered at National Institute of Community Development, Hyderabad (1–33).

Jain, J. K. 1974*b* *Water resources*, pp. 1–18. Government of India, Ministry of Agriculture and Irrigation.

Jain, J. K. 1975 Planning of ground water development for irrigation in India. Report for discussion (Part I) at the Ninth Congress on Irrigation and Drainage, Moscow, 1975, Question 31 (31.233–31.241).

Discussion and replies by the author

B. H. Farmer (*Centre of South Asian Studies, University of Cambridge, Laundress Lane, Cambridge CB*2 1*SD*). May I first congratulate Mr Jain on his interesting and important paper. I was particularly glad that he emphasized the fundamental difference between the groundwater resources of the alluvium on the one hand, and of the crystalline tracts on the other; for I have often been depressed by the ignorance of this distinction shown by academics, administrators and others, who have imagined that the abundant groundwater resources of the Punjab are replicated in other areas of India, and that the Punjab agricultural miracle can be repeated by tapping underground water. In thus underlining the limitations on India's ground water resources, I would, however, like to take the matter a little further than Mr Jain did and to draw a distinction between alluvial areas like the Punjab where groundwater is fed by inflow from large snow- and monsoon-fed rivers and from irrigation canals, and those in the Rajasthan desert where these conditions do not obtain and where the water table over wide areas lies too deep to be tapped.

As for the crystalline tracts, Mr Jain has explained why and how underground water resources are limited and shown why careful investigations are necessary. Work done by our research project, based on the Universities of Cambridge, Madras and Sri Lanka, has shown, further, that in North Arcot District of Tamil Nadu (South India) increases in rice production owe more to electric pumps on open wells (old ones newly dug) than to 'high yielding varieties' (this varies a point made by Mr Jain): but that greatly increased pumping has led to a secular fall in the water table at an average rate, over the last few years, of not less than 0.3 m per annum. In other words, underground water is being used at a rate generally exceeding the recharge rate. Similar results have emerged from the crystalline areas of South and Central India. These facts underline the limitation on the extraction of groundwater and the need for careful investigation and monitoring.

Reply: I agree with Mr Farmer about the distinction between alluvial areas in the Punjab and those in the Rajasthan desert. His remarks about the over draft situation developing in the crystalline areas of the south and Central India and about the need for careful investigations and monitoring in this region are also pertinent.

H. Rush (*Science Policy Research Unit, University of Sussex*). Mr Jain refers to the building of private tubewells among Indian farmers as a means of giving the farmer greater control over the agricultural infrastructure. While the increase in private ground water exploitation has been extremely rapid the rate of well utilization, as reported by I. D. Carruthers, is often extremely low. With the constraints of shortages and high prices for electric power, oil and spare parts for pumps, private farmers operate their wells at the bare minimum and are rarely prepared to sell water to neighbouring farms. Until complementary facilities are improved the constraints to economic utilization of both private and public tubewells will remain and farmer control over the agricultural infrastructure is illusory.

Reply: It is true that in many areas in the country, private and public tubewells are underutilized but private tubewells give a greater confidence to the farmers as regards the assurance of the irrigation supply when they need and this induces them to make higher investment on other complementary inputs necessary for productive irrigated agriculture.

H. W. West (*Wolfson College, Cambridge*). I would like to draw attention to a linkage between arguments put forward this afternoon by Mr Jain concerning groundwater development and mention of the fragmentation of holdings made this morning by Professor Dharma Kumar.

Such fragmentation will introduce diseconomies in groundwater use and hence disincentives working against private investment in the sinking of tube wells, because of difficulties in surface distribution. In such circumstances the need to develop groundwater resources must provide an added motive for the prior consolidation of holdings.

Reply: I fully agree with the remarks of Mr West.

J. A. Allan (*School of Oriental and African Studies, Malet St, London WC1E 7HP*). India is a 'huge parish' as suggested by Professor Kumar earlier today, so I hope that I may be forgiven if I refer to that part of it with which I have some familiarity, namely the districts of western Uttar Pradesh drained by the Ganga–Jamuna river system. This is an area which was affected a century or more ago by the development of canal irrigation, precipitating what some have referred to as 'the first green revolution'. As a result of the canal building water tables rose rapidly with all the attendant problems of waterlogging and soil deterioration.

The recent 'green revolution' has been associated with an even greater extension of water utilization made possible by the massive well drilling programmes of the State Tube Well Organization and especially through the privately developed tube wells of the progressive farming community of western Uttar Pradesh. The aquifers of the deep alluvium of the area are presumably interconnected and clearly have up to now been equal to the massive withdrawals of water witnessed up to 1975. I believe, however, that there is evidence that the water supply in the region of Delhi may not be sufficient to meet the agricultural needs of local farmers. Is this because of the urban demand of the nearby capital or is there some other explanation?

Reply: Groundwater availability in Delhi region is very much limited, because of the presence of hard rock in the sub-strata, as compared to the dynamic reserve of groundwater available in the alluvial areas of Uttar Pradesh.

B. Dasgupta (*Institute of Development Studies, University of Sussex*). I would like to make four points regarding the use of tubewells or irrigation pumps. First of all, their use is dependent on an adequate supply of electricity or diesel fuel. In a country like India, where the indigenous supply of diesel fuel, and the capacity to import are limited, and where the electricity facilities are non-existent in a great majority of villages, the scope for the use of pump irrigation is limited. Secondly, in the case of those villages with electricity, where the supply of electricity comes from hydro-electric projects, the availability of irrigation water largely depends on weather conditions. Thirdly, due to lack of coordination between the electricity and irrigation authorities, irrigation pumps do not receive electricity connections for many months. The conflict of interest between the two authorities lies in the irrigation authorities wanting to space the tubewells in a way which minimises the depletion of groundwater, while the electricity authorities would prefer them to be bunched for reducing costs. Fourthly, tubewells being costly are usually owned by the rich farmers with adequate financial resources, credit-worthiness and access to credit institutions; even when water is sold to poorer farmers, the latter pay a higher price than and remain dependent on the former. A major consequence of tubewell irrigation in Punjab (of both India and Pakistan) is the increasing disparity of income and resources among the farmers.

Reply: In spite of the limitations pointed out by Mr Dasgupta, a programme of groundwater development has been progressing very satisfactorily in India. Public tubewells and private wells owned by groups of farmers are being specially encouraged to meet the needs of the small farmers.

W. H. Boon (*Formerly Plant Protection Ltd, Godalming, Surrey*). We have heard some excellent descriptions about the provision of water supplies. I wonder whether as much attention should be devoted to the utilization of their supplies by modification to traditional cultivation methods. I have in mind, in particular, some work done by my former company in South India which showed that the use of chemically assisted minimum cultivation of rice could lead to a substantial reduction of the water required for crop establishment.

Reply: I completely agree that apart from development of irrigation, the economic use of irrigation water also needs to be given adequate attention.

Phil. Trans. R. Soc. Lond. B. **278**, 525–535 (1977) [525]
Printed in Great Britain

Experience in the Middle East

By V. C. Robertson
Hunting Technical Services Ltd, Boreham Wood

The paper considers the possibilities for meeting the hazards in agriculture – scientific, technological and socio-economic. The discussion is divided broadly into three sections, covering the physical situation; other factors (such as population and the availability of both expertise and funds) which need to be taken into account; and pointers to future improvement.

The first section discusses the resource basis, emphasizing the effect which long-term exploitation by man has already had and indicating, mainly from actual studies undertaken, how evaluation of potential can be undertaken. The second section discusses other constraints which have hitherto tended to hold back agricultural development, and the new impetus which, wisely handled, the vastly-increased oil revenues of the area can supply. Finally, a brief survey is presented covering the directions which future agricultural development in the region might take, and some of the key inputs which may be needed.

1. The physical situation

I want to start with an obvious but fundamental issue. Like China, the Indian sub-continent, and perhaps parts of North Africa, most of the 'Middle East' has been occupied by man, often with substantial densities of population, for a very long time. This makes the Middle East a somewhat special area when we come to consider both its problems and its potential.

It is generally agreed that this long-term use by man (and his domestic livestock) has resulted in widespread degradation or deterioration of the resource base. The deterioration which has taken place is of course of different kinds and degrees. In some areas both vegetation and soil have been almost totally lost; elsewhere land has been permanently degraded by irrigation, particularly in Mesopotamia; the rangelands of parts of Syria, Jordan, Iraq and Iran offer examples ranging from recoverable overuse to complete destruction. Such deterioration may develop slowly over a long period of time – which is almost certainly true in the case of the irrigation in Mesopotamia; in other cases deterioration has been rapid during no more than the past two or three decades. Because both vegetation and soil have been lost, the environment has in effect become more arid in many areas.

The two principal resources with which we are here concerned are land and water. In both cases the *unused* or *under*-used resources are limited, at least at the present level of inputs. Put another way, most *land*, if not water, is over-used at existing levels of inputs. Of the major rivers in the region, the Tigris and Euphrates are in a class of their own, but even their unused waters are limited, especially in summer. Actual and projected reservoirs can and will improve this situation, but whether the available water is being most effectively used is quite another matter. I shall return to this subject later. In Iran, the Karun and Kharkheh rivers probably account for something like one-third of the potentially available surface waters, while the rivers in the extreme northwest, and in the Caspian area, account for more than half the remainder. The Euphrates and, particularly, the Tigris flow through vast alluvial plains where land can readily be commanded by water supplies, but elsewhere the physical relation between land and water is not always so straightforward. In Iran, the Karun and Kharkheh, for example, command substantial areas of alluvial land, but this land all too readily becomes saline – indeed much of it is already saline – and this undoubtedly affects the kind and cost of irrigation development. Otherwise the topography is such that much agricultural land is scattered in relatively small individual areas, so that the development problems are quite different from those of the large alluvial areas.

Groundwater is of course another major source. It is particularly important in Saudi Arabia, and it is significant in many other countries of the region. In Iran the groundwater resource provides about one-tenth of potentially available supplies.

Throughout the region, there are large land areas which produce very little, but most of this land would have little real potential for agriculture even if water supplies were unlimited. These areas include vast stretches of true desert – for example the rocky mountain deserts in Iran, southern Jordan and Saudi Arabia, and the basalt deserts of Jordan. Much of the land of greater potential is in alluvial plains, notably in Syria, Iran and particularly Iraq. Yet even on this land, as I have already indicated, salinity and other hazards present difficulties.

Outside areas whose future turns on irrigated agriculture, there are of course substantial areas where rain-fed agriculture could be improved – the Mediterranean zone of Jordan, for example; the wide sweep of country from Lebanon across Syria and through Iraq from Mosul to Kirkuk; and in Iran, where the area suited to rain-fed agriculture amounts to some 6.5×10^6 ha, compared with 5.5×10^6 ha of land supplied with irrigation water, of which about 3.6×10^6 ha are cropped annually. Little of this rain-fed farming area is managed as it could be, though there is a remarkably well-managed and largely rain-fed agriculture, involving complex and well-maintained terraced fields, in the mountains of the Yemen Arab Republic – something I have, to my regret, not yet seen.

Traditional agricultural methods have relied on what has been described as the 'minimal risk' system, applied even in irrigation in the absence of assured supplies. But however sound such systems may be ecologically, they tend to break down under population pressures and the land then suffers. There are also worrying instances of unwise 'land-mining' operations, using machinery and financed by the now readier supplies of capital. The ploughing of marginal rainfall or 'steppe-zone' rangelands (around 250 mm rainfall) for short-term catch crops of wheat is perhaps the worst example. Once rangeland vegetation, however degraded, is destroyed completely in this way, it is very difficult indeed to re-establish.

One has to conclude that, in sum, the basic resource situation is not too encouraging. There are of course bright spots, and it is certain that very major increases in agricultural production could be achieved through proper management of the resources now available. But major efforts in conservation will be needed, and the social problems involved, in shifting or settling people, or in reducing stocking rates, are very real and difficult to deal with.

2. Some other relevant factors

Population pressure seems bound to increase. In most countries of the region populations seem likely to continue to expand at rates between $2\frac{1}{2}$ and 3 % per annum. At the same time, particularly in those countries with oil revenues, rising incomes will not only generate increased demand for food but are likely to alter the pattern of demand – perhaps especially in relation to livestock products. But the trend towards urbanization and industrialization will perhaps begin to ease the direct pressure on land and so make it easier to introduce more rational and eventually more productive methods of land use.

Many countries will probably try to become as self-sufficient as they can in foodstuffs. This is entirely understandable, yet it will make it more difficult either to ease the pressure on overstrained resources or to rationalize production so that it makes best use of the resources of the region as a whole. One would like to feel that the new Arab Fund/U.N.D.P. study on Inter-Arab Integration and Development, due to have started this year, could point the way. This study of course extends beyond the Middle East; a good deal of attention is being directed to Sudan as a future source of food for the Arab world.

The present average yields of crops are small, and have changed all too little over recent years – though there are some encouraging exceptions. The reasons why yields remain small are complex, but they include degraded land, insufficient or untimely water supplies, lack of improved agricultural inputs, uncertainties caused by land reform, and chronic shortages of trained manpower and management. Although they differ in degree in different countries, these problems remain largely unsolved.

Land reform is such a crucial – and politically sensitive – issue that it needs separate comment. In many parts of the Middle East, the pressures for land reform have been great. All too often, however, reforms have been introduced without any understanding of how the old system, however bad it was, actually worked. As a result exploitation of people has sometimes been replaced by even greater exploitation of the land, simply because the old methods of management have been removed – along with the agricultural inputs which they sometimes included.

The main new factor, which may be crucial if it is wisely handled, is the sudden increase in wealth flowing from the increased oil prices. Although the direct effects on individual countries obviously differ widely, the indirect effects are or will be widespread. Properly used, this new wealth offers a very real opportunity to steer agriculture out of its present relatively static state and, in the longer term, to stabilize and renew the resource bases. At the present time there is sadly little coordinated planning, and significant amounts of the new wealth tend to go into wildcat or ill-considered schemes – including the purchase of 'the latest technological developments' which have important prestige value but do not necessarily offer a practical path to development under present socio-economic conditions. The latest technology is not necessarily the most appropriate.

3. Analysing potentialities

The two examples I want to give here draw on experience in Jordan, Iraq and Iran. The first concerns the broad appraisal of land-use potential – or appropriate land use – on a regional or national scale. The second concerns analysis of a particular form of development, in this case irrigated agriculture.

Land-use potential

In appraising regional land-use potential, the first step is to develop what I call 'zoning'. Zoning of land, in this context, establishes what are the 'proper' uses, or range of uses, for different land units, irrespective of the actual uses to which they are now being put. Proper (or optimum) land use describes ways of using land which will permit sustained production without physical deterioration of the land. In my view, proper land use includes the preservation of amenities (such as complete protection of areas of great scenic beauty), the protection and management of wild-life in national parks, and the protection of areas of specific scientific interest.

Most countries in the region already possess enough data to develop at least the first stages of a zoning system. Data on climate and soils are of the first importance, plus information on landform and topography. To this one should certainly add present land use (including information on any remaining natural vegetation such as forest and rangeland), since so much of the effort for optimal use of land lies in redirecting or improving the uses to which people put it already. Most areas have substantial soils information: indeed though a great deal of time and money has been spent on soil surveys, surprisingly little of the information from them has actually been used for planning or land management. It is often claimed that attempts to develop a comprehensive zoning system take too long and are too expensive to consider. In fact most of the necessary data are usually available; commonly the problem is that no effort has been made to coordinate or synthesize it for the purpose of making land-use decisions. The wide availability of aerial photography, now supplemented over virtually the whole of the Middle East region by 'Earth Resources Technology Satellite' (ERTS) imagery, also means that a mass of 'remotely sensed' data is available to the planner at minimal expense.

A useful first step is to divide the country into areas which are uniform in landform and vegetation. The types thus distinguished, and the maps of their distribution, can be used in more detailed investigations. We did this in Jordan, though the immediate objective – a 'range classification' of the natural pastures – was much more limited. The survey was however designed in such a way that it could be used as a general land classification which could later be applied to other forms of land use. The survey was supplemented later by work on the bioclimatology and vegetation of Eastern Jordan by G. A. Long, an ecologist from Montpellier: the two studies combined provide the essential basis for comprehensive land-use planning.

The survey covered all of Jordan east of the Jordan River – the greater part of the country. The climate ranges from sub-humid Mediterranean to arid Mediterranean, with large areas of desert classified as Saharan bioclimate, from cool to very warm. (The terminology is that of Emberger.) Climate critically influences land use, but physical conditions have been markedly degraded in the areas more favoured for human occupation. In some areas this degradation has developed over a long period: elsewhere it is relatively recent. For example considerable areas of the remaining oak forest in the south-central highlands were destroyed when the Hejaz railway was built, around 1916, and as a result much of the soil cover was lost.

The survey work in Jordan covered nearly 97000 km^2. The team's work in the field was concentrated on collection of sufficient information to distinguish and describe individual ecosystems or biological habitat types; and to identify these, so far as possible, with characteristics recognizable on the aerial photographs. The results could then be extrapolated to cover the whole survey area. Because little time was available for field work it was rarely possible to return to any area, so that all necessary information had to be collected during a single visit. For the same reason, the examination of soils could be no more than elementary. In an area where little basic information on soils or vegetation was available, the opportunity to have a second look can be important for many reasons. It is now clear that these were disadvantages which should be avoided in future studies of this kind. As a general rule, vegetation is one of the most useful indicators of human influences on local conditions. However, vegetation which has been used by man for very long periods, as is the case in Jordan, has usually been so deeply modified that it reflects present conditions only. Human interference, in semi-arid or arid regions, nearly always changes vegetation so that the environment becomes even more arid, but relict species may provide evidence of potential. A fully useful classification system should consider not only the present state of the vegetation but the capacity for improvement.

In Jordan we were able to combine a bioclimatic classification with a mapped distribution of ecosystems or habitat types. This had two advantages: in the first place, effective cross-checking became possible and, secondly, the potentialities in the various bioclimatic sub-divisions could also be more fully identified since the level of physical resource degradation – where it had taken place – was also known. In the 'sub-humid Mediterranean' bioclimate region of Jordan, for example, it is still possible to find areas of something approaching climax forest, with soil cover intact. Elsewhere, within the same bioclimatic region, almost all vegetation has disappeared and little more than a bare rock surface, denuded of soil, remains. Unless situations of this kind are recorded, as they were during the survey under discussion, the usefulness of a bioclimatic map, on its own, must be limited.

Since the Jordan survey was done, nearly 20 years have passed. A great deal of additional resource and climatic data has been acquired throughout the region. Moreover, the ERTS satellites, with virtually total and recurrent cover in both monochrome and false colour, provide a wealth of information – admittedly at a very small scale – which can be used to analyse landforms and related features. No country concerned with the long-term future of its land resources and, therefore, with the well-being of its people, should fail to make use of this outstanding source of data. Supplementary work may of course be needed in varying degrees, but none of it is costly in terms of the issues at stake.

The same kind of approach was used later in studying the Diyala river basin in Iraq. The lowest parts consist of a level plain of alluvium, where irrigated agriculture was clearly the most important way of using land. It was in the upper and middle parts of the basin, where choice was widest, that the kind of study under discussion was most appropriate. The stated objectives of the study were clear, and again somewhat limited. They were directed essentially to three questions: the extent of present or potential erosion in the watershed area; the nature of present forest or other vegetation cover and its management in relation to erosion control; and a study of existing agriculture and measures to be taken both to raise production and to reduce erosion. Nevertheless the main elements of comprehensive land-use planning were there.

As a first step, the area was divided into ecological zones based on vegetation characteristics. The zones were designated dry steppe, moist steppe and established forest. The most important ecological boundary was that between dry steppe and moist steppe, two plant formations which differ greatly in their productivity and botanical composition. In moist steppe, rainfall is sufficient to maintain a continuous cover of trees, though most of these have gone as a result of human activities. In true dry steppe the climax vegetation is grassland, since the rainfall is insufficient for trees. The division between these two kinds of steppe approximates to the 300 mm rainfall isohyet, and this was probably where, originally, forest passed into steppe, through a wide zone of steppe-margin woodland. Most of the remaining established forest belongs to a dry type of the Iraq mountain forest characterized by species such as *Quercus aegilops* and *Pistacia khinjuk*. The climate under which forest exists – and would have existed – in the upper Diyala is exceptionally extreme (very hot summers and cold winters). Under pressure, in such a climate, ground conditions change drastically in the direction of greater aridity, and an invasion of steppe species into the degraded forest is likely. Once the forest is removed it is very difficult to re-establish: in this region degradation from woodland proceeds straight to annual grassland without the shrub stages which intervene in the Mediterranean region.

The broad ecological zones initially determined were further subdivided according to landform and geological formation. These divisions were very significant. In the moist steppe zone, for example, most of the productive grassland was on conglomerate rocks. The soil, though often not deep, is remarkably stable and resistant to erosion, and hence the conglomerate grasslands can withstand heavy grazing pressures. Nevertheless, the effects of protection from grazing are very marked: perennial grasses became established quite rapidly and, given a source of seed, oak trees would probably reappear. The geological formation known as Fars provides a striking contrast to this relatively stable situation. The Fars beds consist of interbedded sandstones and soft clays. The landscape is composed of bare sandstone ridges and red clay badlands, whereas the conglomerates produce a gentler landscape of rolling rounded hills. Within the Fars formation three landforms can be distinguished: thin soils on the exposed sandstone, clay badlands, and valley bottoms (often quite extensive) with deep alluvial/colluvial soils. The effects of human use, and the potentials for use, are therefore widely different in these two geological formations, although both may occur in the same ecological zone. By analysing and mapping these factors, the necessary information for sound land allocation becomes available.

In Iran, in a broader context altogether, we were particularly concerned with defining the limits for rain-fed agriculture, which could again be related to certain climax forest vegetation formations: Alborz deciduous forest along the Caspian shore and Zagros deciduous forest (an oak–pistachio woodland) for what we have termed 'high-potential' rain-fed farming with average rainfall around 450 mm or more, and steppe forest (a *Pistacia–Amygdalus–Juniperus* woodland) for 'low-potential' rain-fed farming (250–450 mm) or improved range. Selected field traverses were made in an attempt to establish these quite critical zones at the small scale necessary for national-level planning. Subsistence or catch-crop arable farming below the 250 mm rainfall level should be banned and the land returned to managed rangeland use.

Irrigation

Turning to irrigated agriculture, I want to consider the situation in the Tigris/Euphrates plains of Iraq. I have selected this area for two reasons. First, these plains perhaps represent the largest 'high-potential' area in the Middle East. There is also an extremely long tradition of irrigation which has clearly had a major effect on the situation we see today. Not only have substantial amounts of salt accumulated in the soil, which is not surprising under long-continued irrigation without drainage, but in addition the soil structure seems to have deteriorated, which makes irrigation difficult because infiltration and drainage are slow.

The evidence of past use is so clear (the networks of abandoned canals of various periods can be seen on aerial photographs), and the salinity/structure problems are so widespread, that we began to speculate whether the development of salinity in the soil had been a factor in the rise and fall of the ancient city states of Mesopotamia. An archaeological expedition from the University of Chicago became keenly interested in this concept, but the results were inconclusive: a canal system is far more difficult to date than an established settlement.

Salt in soil can be removed by leaching and associated procedures, but the degraded structure is more difficult to correct. Its nature remains something of a puzzle, though it seems clearly related to the old irrigation systems. Extreme cases show a curious surface relief similar to the Australian 'gilgai', which gives a very characteristic pattern on air photographs. We originally described these soils as 'solonetzic', but the exchangeable sodium values give only very slight support for this classification. It seems that deteriorated structure is not related to any major difference in exchangeable cations. Nor is it wholly a manifestation of dispersion (in the sense that a sodium-dominated clay disperses to give a bad structure). It may be due to a closer packing of soil particles leading to a structure with smaller pores. (Laboratory measurements relating suction pressure to soil water content tend to support this view.)

Whatever the solution, the problem does affect the selection of areas for development and the pattern and intensity of irrigation supply. Almost certainly areas of extremely poor structure should be excluded, and fortunately they are very easy to identify. (Apart from their distinct pattern on the aerial photographs, we found it instructive that local people employed as pit-diggers and auger-borers consistently identified such soils as unusable.) The tendency, hitherto, without a drainage system, has been to spread irrigation supplies very thinly over very wide areas, with cropping intensities as low as 50 % in winter and 10–25 % in summer. It now seems wiser to concentrate on the better land, installing drainage and adopting a more intensive cropping pattern, hoping thereby to lessen salinity, improve structure and greatly increase crop yields. The proof of this particular pudding will of course be in the eating, and most of the meal is yet to come. If this policy succeeds, the consequences will be widespread and important.

4. Pointers to future improvement

It seems inevitable that irrigation must provide the main bases of improved agriculture in the region, simply because of the generally arid situation throughout most of it. Yet, in spite of a wealth of studies and considerable investment, development (in terms of crop production rather than of completed engineering works) has been all too slow, largely because post-construction management, extension, farm-level inputs, regulation of the prices of farm commodities and the establishment of marketing facilities have received little attention. Nor has anything been done about rural infrastructure and amenities. None of these things present insuperable difficulties once the need for them is accepted, and funds are generally available. What can be done, in a harsh environment, with technical ability and skilled management, has been clearly demonstrated in Israel: but political relations have prevented any direct transfer of such success, at least to the Arab countries.

In addition to, or in conjunction with, irrigation, major development in livestock production is possible, provided livestock are integrated with agriculture so that resources are used (and conserved) in a rational way. Rangelands in marginal rainfall zones can be managed as breeding areas, particularly for small stock, and the animals can be finished in irrigated areas or on rain-fed farms in wetter regions. Both Iran and Iraq are beginning to develop large-scale commercial enterprises, mainly so far in dairying and poultry. Unless some reorganization of this kind takes place, the rangelands will continue to deteriorate. In Iran they constitute a vast resource whose effective management is at present largely neglected. In Iraq, Syria, Jordan and Yemen they are less important but nevertheless significant. To rationalize livestock production requires major social changes, including changes in political attitudes towards nomads and in the concept of management (including control of stock numbers) among the nomads themselves. No country has yet, so far as I know, evolved a coherent policy in this respect. The tendency is to shy away from such problems, and often to condemn nomadism as primitive without fully appreciating it. In Iran, the rangelands vary greatly in altitude and it is necessary to organize some kind of seasonal rather than year-round use. A nomad or at least transhumant system is still appropriate, provided stocking controls and proper stock routes can be introduced.

Catchment protection is needed almost everywhere in connection with surface storage of water, but it is largely neglected except perhaps in Iran. The upper river catchment areas usually receive more rainfall and provide opportunities for afforestation (and hence timber products) and other tree crops (fruits and nuts) instead of the progressive deforestation and denudation that continues almost everywhere.

Though I have commented critically on *new technology*, it must not be dismissed. But its application and particularly its transferability need careful examination. The use of oil or gas to produce animal feedstuffs could be important, not only to increase livestock production but perhaps also to take pressures off rangelands temporarily so that conservation and improvement could be initiated. A breakthrough in desalination of water, so that its cost becomes compatible with irrigated agriculture, could be of vast importance. Even now, small-scale plants are operating in Kuwait and Abu Dhabi, producing valuable vegetable crops in what is virtually a hydroponic system, and in at least one case, using surplus energy (from gas which would otherwise be burned off).

Then – a vital factor – there is a widespread need for *training* of manpower at all levels, but perhaps critically in upper and middle management. The shortage of managers, perhaps more than anything else, cripples decision-making and the successful implementation of development schemes, as the countries of the region are increasingly recognizing. How to provide such training is a complex question, yet it is extremely important to find an answer.

I have spoken mainly about the evaluation of technical problems and solutions. Of course any worthwhile planning must also consider the economics of different courses of action, and pay proper attention to sociological factors – after all farming is undertaken by people and there are plenty of examples about to illustrate what happens if this is ignored. As a recent article in *The Ecologist*† put it: 'Scientists and development planners work out elaborate schemes for rural regeneration, but peasants and goats seldom seem to find it in their own interest to assume the profile of the computer cards they are dealt.' But although it is possible to see rather clearly what is happening and what should be done, it remains difficult to bring about the appropriate action. It is difficult to get the decision-makers to see the problems in the round: to provide for effective management and farm-level inputs on an irrigation scheme rather than merely complete the engineering works; easy to get decisions to build a dam and very difficult to get anyone to pay more than lip-service to watershed management; and even more difficult to get those in power to consider rationalizing land use for the long-term productive management of their country's resources. One is faced constantly with the old problem of short-term expediency in preference to long-term benefits. One can understand the difficulties the decision-makers face in such choices, while remaining seriously concerned about what will happen unless they are faced. There seems little real awareness of the

† E. Eckholm 1976. The politics of soil conservation, **6**, no. 2, February 1976.

importance of conservation – and I use the term in a positive sense, implying sound resource management and sustained production – and without this it is difficult to see how even the best-conceived plans can be put into effect.

Fundamentally, countries in the Middle East, like many other countries, must devote a great deal more attention, and a greater share of their budgets, to agriculture – a fairly basic change in thinking and priorities.

In a part of the world that still lives to a great extent on the exploitation, the mining, of non-renewable natural resources, such basic changes in attitude are slow in occurring. But I think they *are* taking place, and that is perhaps the most hopeful note on which I can end.

Discussion

T. Riney (*Department of Forestry and Natural Resources, Kings Buildings, University of Edinburgh*). In commenting on Mr Robertson's paper I would like to mention two points. In 1972, in response to a request from the F.A.O. regional office in Cairo, a three-man team, comprising a sociologist, economist and ecologist, visited Iraq, Iran, Sudan and Libya to identify the marginal land problems in each of these countries. These countries were selected by the Cairo office as being representative of the region. In terms of an analysis of the environmental problems our findings were essentially the same as described by Mr Robertson. It may be of interest to know that, although we recognized the core of the problem as a man-mismanagement-environment problem, we gave priority to a social constraint: lack of coordination of aims and activities existing between different government departments. Working with responsible officers in each country, this difficulty was suggested by the officers themselves as the main bottleneck to progressing towards effective stabilization of their marginal lands and toward eventually developing more appropriate forms of use or management. Two countries added the additional qualification 'of course before we can coordinate we must first communicate' and communication between departments was very poor indeed. We recommended that, in considering future assistance to the region, solving this sociological constraint should have high priority: either as a prerequisite to other normal kinds of assistance; or as an integral part of future projects and programmes, that is to say building into future projects special mechanisms for achieving such coordination as would be prerequisite to making the project a success.

The second observation involves recognition of the extent nomadic traditions are long-standing and widespread and the possible significance of comparatively recent breaks in these traditions. In Iraq of special interest to us was the pattern of utilization by pastoralists. There a traditional nomadic and semi-nomadic pattern had continued for 3000 years, this pattern having spread and being essentially the same as shown by nomadic pastoralists in the Sahara. It seems important to understand on the one hand the reasons for the down-grading trends often associated with such changes as has been mentioned by other speakers. On the other hand, more infrequently but equally important is understanding why the maintenance of a stable environment has been possible in spite of rather sudden breaks with tradition. It seems to me that using a 'uses and limitations', or 'dangers and benefits' approach to research studies directed at understanding breaks in the nomadic tradition and their consequences, could provide the kind of basis of understanding which might accelerate the process of profiting from past experience – a process apparently little appreciated in developing resources in semi-arid lands.

D. C. P. THALEN (*International Institute for Aerial Survey and Earth Sciences* (*I.T.C.*), *Enschede, The Netherlands*). A remark has been made in the discussion that the use of the Middle Eastern Rangelands has not changed over the last 3000 years. I have to disagree, especially taking the last 30 years in comparison to the long period before. My information is based on a few years field work in the desert areas of Iraq and on literature data. A major change in the use of the arid and semi-arid ranges has taken place over the last decade, due to new mechanical means allowing the drilling of wells and reaching remote areas with 4-wheel-drive terrain cars. This change has caused an enormous increase in grazing pressure. In this respect a parallel can be drawn with Mr Perry's picture of the Australian rangelands following the settlement. Statistical data on the total number of sheep in Iraq may illustrate this. The first data are from the thirties giving figures of about 4–5 million. This is followed by an increase to about 10–11 million in the mid-sixties. These figures have in some publications been extrapolated to about 14 million in 1973. Recent, in my opinion reliable, data show that the figure for 1971 should be around 6 million. This fits in Mr Perry's picture for Australia of an increase in livestock number when opening up new areas, followed by a sudden drop when the resources had been over-used and a gradual stabilization afterwards at a lower level under proper management of the already degraded resources.

In my opinion in the Middle Eastern rangelands we have gone through the first two phases in the last 30 years or so. It is time to be concerned about the third phase, that of proper management based on survey and research data.

J. A. ALLAN (*School of Oriental and African Studies, Malet Street, London WC1E 7HP*). It has been suggested by Mr Robertson that man has mismanaged water and soil in such a way that the soil has become unusable because of saline–alkali degradation. Professor Jolly has further enquired whether the process can be reversed so that degraded areas can be reclaimed. I am not as familiar with the Tigris–Euphrates area and its irrigation systems as I should like to be, but I should be surprised if the position there is not similar to that which I have encountered in other semi-arid areas in northwest India. There the climate and geomorphology are such that the development of saline–alkali soils is a 'naturally' occurring phenomenon, and it is likely to be a condition which has affected progressively more and more of the area during past millennia. Could I enquire whether the irrigated lowlands of Iraq show evidence of similar trends which irrigating man should recognize in managing water, and how this might affect approaches to soil reclamation in such areas?

V. C. ROBERTSON. It is probably true that development of saline-alkali soils is a naturally occurring phenomenon in a situation like the Tigris–Euphrates plains, where drainage is naturally poor and salts are distributed at times of flood. It would be a very slow process, however. 'Irrigating man' can, however, greatly speed this process by increasing the application of salt in irrigation water and by building up the level of groundwater which may itself be saline. These factors must be taken into account in designing irrigation schemes, so that salt build-up is prevented (by use of suitable 'leaching' factors) and groundwater levels controlled by properly designed drainage systems.

E. B. WORTHINGTON (*I.B.P., c/o The Linnean Society, Burlington House, Piccadilly, London W1V 0LQ*). With reference to the extensive ancient irrigated areas of Iraq, deserted through soil salinization, and the idea that these areas had been used successively on a system which might be compared with shifting cultivation, was there evidence that the salinated irrigated areas, like areas of rain-fed agriculture, recovered within measurable time?

V. C. Robertson. We obtained no evidence that salinized irrigated areas could recover, though theoretically some recovery is possible if cessation of irrigation allows high water-tables to drop and salt to be leached out of the upper soil by the limited rainfall. There is some direct evidence that such recovery occurs in the Indus plains of Pakistan.

Phil. Trans. R. Soc. Lond. B. **278**, 537–542 (1977) [537]
Printed in Great Britain

The evaluation and exploitation of the West African Sahel

By J. Dresch
Union Géographique Internationale, Institut de Géographie, Paris, France

A critical review and discussion of: (*a*) the natural resources of the Sahelian Zone in West Africa and their use by the population to strike a balance between agriculture and pastoralism; (*b*) the incidence of droughts; (*c*) the results of economic and social changes during colonial times and since independence; and (*d*) the solutions that have been proposed for the problems that have arisen.

The West African semi-arid lands fringe the Sahara to the south. This is why it is called in arabic Sahel. It is a transitional region between desert and savanna. Rainfall increases gradually towards the south; in the northern Sahel rainfall is between 100 and 300 mm, falling during the 1.5 to 2–2.5 monthly wet season; agriculture is not possible. In the southern Sahel rainfall is between 300 and 500–600 mm or somewhat more, the rainy season lasts between 3 and 4 months; agriculture as well as stock breeding are possible. But rainfall is irregular, stockbreeders and farmers are threatened by droughts, and famines are frequent so that, between 1968 and 1973, international assistance has been necessary. The people are poor, among the poorest in the world, and solutions to development problems and even to the maintenance of population and of human activities in this area have yet to be found.

1. The natural resources

Resources are rather similar from west to east, from Mauritania and Senegal to Tchad and Sudan through Mali, Upper Volta and Niger, because Africa is a flat land, all along the Sahel, without any mountains, or wide ranges of altitude between plateaus, plains and valleys. There are, of course, varying natural conditions between rocky plateaus and their escarpments, some isolated rocky hills and large alluvial plains and valleys. They are often covered by sands, ancient dunes inherited from dry periods when, during the Upper Pleistocene, the desert advanced several hundreds kilometres southward. Those Sahelian sands are rather good soils because they store the rain, pools hold water between the dunes. The alluvial plains and deltas, more or less flooded during summer and autumn, frequently provide valuable soils and moisture when rivers coming from the southern more rainy zones bring an additional supply of surface and underground water. Elsewhere, the ferrugineous tropical soils are generally poor.

Though the geomorphological features differ somewhat from place to place, the Sahel looks like a rather monotonous landscape because vegetation does not show large differences along the belts extending from west to east. It is called a steppe, the main patterns of which are trees, deciduous trees, mainly acacias at least within the driest belts, and various Graminaceae or other families form an open formation. This steppe is easily cleared for cultivation and does not burn as savanna grasslands. So the Sahel can be used either by Saharian nomads, such as the Tuaregs coming during the dry season for grazing, mainly within the sandy areas or along the pools and rivers, or by cattle breeders such as the Fula, or by farmers who can be stockbreeders as well.

However, all of them have to be cautious because rainfall is irregular. First of all within the year so far as the summer rainy season is short, begins more or less early, ends more or less late, the first showers are followed by too many dry days. So a rainfall is efficient if it is more than 3 mm and is followed by another fall less than a week later. The irregularity is evident also from one year to another. The lower the rainfall, the greater the irregularity: the ranges between maxima and minima rise from 2 to 5 in southern Sahel, 1 to 5 or 10 in northern Sahel. So average rainfall has no significance. Dry years can follow one another. Several droughts have been noticed and people remember famines. The last ones occurred between 1909–16, 1940–8, 1968–73, but their importance and their length were different from one region to another. Between the droughts, a sequence of years may be humid, for example between 1960 and 1968, so that one isohyet may move several degrees, up to 5, in latitude. The inhabitants of the Sahel know those irregularities quite well. The question for us, people of the developed countries, is

(1) whether there are regular sequences of dry or humid periods and therefore whether we can correlate them with other meteorological and geophysical data and forecast disasters;

(2) whether there is a general tendency towards an increase of aridity or not.

Unfortunately it is evident that up to now we are unable to forecast years or periods or to certify that there is a long-term increase or aridity: it may be shown in Mauritania, but it is not evident elsewhere.

We can conclude that the Sahel is not lacking in resources. Different kinds of peoples can find either pastures for camels, cattle, sheep and goats, or land for cultivation. But there is a determinate relationship between the livestock, the surface under cultivation, the number of people, their production modes, their standard of living and the governments' policies. Within the countries of Mauretania, Senegal, Mali, Upper Volta, Niger and Tchad, there are a total of around 24 million inhabitants, but a small proportion lives in the Sahara, a larger in savanna areas. Maybe 7 million are living within the Sahel. The g.n.p. in those countries is estimated at US $ 40–50 in Upper Volta and Mali, and a maximum of $180 in Senegal. But it is evident that in the Sahel it is lower than the national average. Therefore, the question is whether land use in the Sahel was – and is – able to provide subsistence to the inhabitants, whether the standard of life could be raised with better integration into the national community and whether and why there is desertification, in so far as there is no evidence of a general increase of aridity.

2. The exploitation of natural resources

Stockbreeders as well as farmers know rather well how to use the Sahel resources. The pastoralists, either in Mauretania or Tuareg further East, in Sahara and Northern Sahel, or Fula, the cattle breeders of the Sahel, move with their livestock from pasture to pasture, varying according to seasons, years, types of livestock, or agreements with other peoples. They gain security by those movements, by spreading their livestock, raising the number of their animals, and by diversification. When they were victims of droughts, they lost, of course, part of their livestock. Families, mainly children, could die, but when rainfall increased again, the community was able to restore the balance between natural resources and their exploitation. Besides, the Saharian pastoralists have long been politically dominant in the Sahel, from the Middle Ages and the time of the great empires of Ghana, Mali and Gao. Owners of camels and horses in the Sahel, they controlled the regional and international routes and the trade of salt, cereals and gold. During the 19th century, the Sahelian cattle-breeding Fula, who became Moslems and expanded Islam, have created several powerful states.

The farmers cultivating millet in the southern Sahel are strongly organized villagers who use to shift their fields according to climatic fluctuations and soils. They store cereals in granaries, so carefully that they can often preserve food and seeds for several years.

Colonization and the creation of independent states have involved important changes in the economic, social and demographic structures of the Sahelian peoples. They can be summarized as follows:

(1) As everywhere in developing countries, the population increases. What is the rate in the Sahel, we do not know exactly. Some authors say that the rate of increase is higher among farmers than among pastoralists. Maybe that is so in general but not everywhere. Mortality rates are certainly decreasing, especially among children, except during droughts.

(2) Social and political structures have been and are being modified. The domination by Saharian nomad or by Sahelian Fula has come to an end. Most of the governments of the independent states are trying to integrate nomad pastoralists into the national society and economy and to promote sedentarization. The strong organization of the society among nomad communities is progressively broken down, the lower classes, or castes, slaves or captives, have been released. They are migrating southward and settling in the Sahel, where they can combine pastoralism and agriculture. The Sahelian Fula are free either to lead their cattle northward during summer, mainly for salt cure, or to dispute the pastures during the dry season with the Saharians, or to settle and combine cattle-breeding and cultivation. The peasant communities are modified also. They are extending cultivation, especially during humid periods, following the rain northward, and on sand soils, towards the limit of 300 mm. The average yield of millet is between 500 and 900 kg per hectare and the consumption is between 200 and 300 kg per year and person, giving between 6.7 and 8.4 MJ (1600 and 2000 kcal) per day. Rice is used together with millet and some milk or meat. They are often indeed either associated with Fula, or become cattle-breeders themselves. They also have sheep and goats, goats being able, better than any other kind of livestock, to find by themselves pasture and to survive during droughts. That is why, in some regions of the Sahel, settled people own more livestock than pastoralists.

(3) The result of those social and economic changes is that Sahelian land is more and more used and settled, especially when humid periods occur. The Saharian nomads cannot desertify the desert by overgrazing because it would involve suicide. In the Sahel, the vegetation cover and the variety of pasture are greater. On good pastures in deltas, or alluvial valleys with pools, cattle breeding needs 3 or 4 ha per animal, and one family can find its subsistence with an average of 50 cattle, often less. As concerns cultivation 2 or 3 ha can provide the cereal supply for one family. In the past, pastoralists and peasants have often quarrelled for either pasture or cultivated fields. In Niger, a limit has been drawn by the administration between nomads and settled peasants. Today, the quarrels may go on, but the opposition between nomads and farmers is slowly decreasing. Nevertheless, their techniques are not modified, they work mainly for self subsistence and do not much increase their participation in the national economy. Why?

During the colonial period and today as well, the Sahel has received much less equipment than the more humid areas to the south. Ground nuts are cultivated further south where rainfall exceeds 400–500 mm, and cotton further on where rainfall is 600–800 mm. In these areas roads have been built and then schools, and some hospitals, etc. This is why ground-nut and cotton belts have extended so much. The Sahalien regions can only provide the more southern areas grain or meat, and the cattle have to walk hundreds of kilometres to the abattoirs of the towns.

(4) The general result is that the Sahelian balance between resources and population breaks down. During droughts, overstocking and overgrazing become dangerous as also does the extension of cultivated land. During the last drought, herds have been decimated, mainly cattle, though many Fula have migrated early towards the south. There were not stores within the granaries for everybody. Pastoralists as well as farmers had to migrate toward the south and the towns. The natural resources themselves cannot be preserved. Sand dunes revive, soil erosion is becoming dangerous, mainly around pools and wells that have been sunk more frequently but where more numerous livestock gather and destroy vegetation and soils. Some estimates have been made of soil erosion: 10–20 tonnes/ha. It may be so in some regions and in dry years, for instance in Northern Senegal where groundnuts have been cultivated too far north. Fortunately up to now the vegetation cover is often quickly restored except for the trees, but what will occur in the future? And what could be the solutions?

3. Solutions

We cannot improve the climate. During the last drought, international assistance, unfortunately too late, has provided cereals so that famine has been stopped without too many casualties. But international assistance does not provide solutions to the Sahelian problems.

New techniques could improve pastoral economy, by enriching the quality of pastures and fallow lands, combining nomadic movements with periodic stops in irrigated lands, improving breed quality, increasing the number of pools, natural or artificial, and of wells, organizing meat or dairy marketing and building abattoirs and other factories within the Sahel and specializing in the Sahel in the breeding of young stock sent for fattening to more humid savanna regions.

New techniques could improve agriculture as well. So far as farmers own more cattle, the use of animal power could extend the area under cultivation and the animal manure could be used as fertilizer. Irrigation could be extended by flood retreat agriculture, by boring deep wells where there are large underground water resources, by the construction of numerous dams to store summer water within depressions. On the big rivers coming from the south down to the Sahel, Senegal, Niger, Chari, more important dams and canals can be constructed, as has been done in the past on the Niger, without indeed convincing success. Better methods could improve the storing of the harvest and the marketing.

So we may propose many kinds of management. Everybody knows that it needs equipment, investment. Furthermore, it needs not only a government policy, international agreements, and international cooperation, but also deep economic and social change. The unstable balance between natural resources and land use has to be preserved. Overgrazing and overclearing for cultivation have to be controlled, some areas protected, the use of pastures, wells, and rivers, regulated. Without the agreement of the populations, and their association with this new policy, nothing will be possible. Agreement and association will be obtained only if young people go to school. But if roads and schools are built within the Sahel, will the young generations stay cattle breeding and millet cultivating? During the last drought, most of the Sahelian population migrated toward the south and into the towns. When rainfall increased again they came back in their pastures and villages. But the young generation have new needs and they are less enthusiastic to go on working and living according to tradition with an income of U.S. $ 100. But what kind of employment will they find in the humid savanna, or forest, or elsewhere?

Discussion

V. C. Robertson (*Hunting Technical Services Limited, Elstree Way, Borehamwood, Herts.*). I would like to raise two points concerning the western Sudan, an area which has not otherwise been touched on.

First, at the time of the drought years in the early 1970s, we heard much of the terrible losses of livestock and of human starvation across the West African Sahel zone virtually all the way across to the Chad–Sudan border. Then such reports reappeared in Ethiopia. One might have thought that Sudan escaped the drought, which of course it did not. The people of Sudan's western savannas were however able to escape the worst effects of the drought years by moving south, which they were able to do firstly because they had no international frontier to cross in doing so and secondly because the wetter country they moved into was not already extensively occupied. This ability to move, with their stock, was often not available to peoples in western Africa.

Secondly, we have for some time been studying the situation of the cattle-owning peoples in western Sudan. Although we are convinced that the raising – particularly the breeding – of cattle is one of the more appropriate forms of land use in this region, we are also convinced that – both because of the relation of livestock numbers to available range and, perhaps more significantly, because of the changing ratio of human and animal numbers – uncontrolled nomadism is not the long-term answer. Indeed the nomads themselves seem to recognize this by settling spontaneously. Building on this, we have evolved an approach to development which involves a rationalization of land use, a separation of settled farming and pastoral use (at least so far as breeding herds are concerned) with the aim of establishing a stable productive system. A first scheme of this kind is now going ahead, with the support both of Government and the local people. Its results could be of very great importance.

R. Dodoo Jr (*Council for Scientific and Industrial Research, Accra, Ghana*). The Council for Scientific and Industrial Research in Ghana has developed a research strategy for our northeast savanna area that is centred round five problem-solving themes, namely:

(1) The burning component: a better scientific understanding is required of the effects of burning on crop production, range and soil resources, and the general floral and faunal ecology of the area.

(2) Cultivation practices component: here it is intended to focus on a thorough investigation of all the cultivation methods now used or potentially employed, as they relate to land degradation.

(3) Water availability and distribution component: information under this head would contribute to the evaluation of the other resources of the region, and to management decisions as they affect development.

(4) Overgrazing and deforestation component: it is imperative to conduct research on the

problems associated with overgrazing and deforestation as they relate to the management and development of semi-arid ecosystems.

(5) Socio-economic research component: long-term solutions to such problems of the region as burning, overgrazing, malnutrition and low productivity need to be found within the context of the region's social, economic and cultural conditions.

W. E. Ormerod (*Department of Medical Protozoology, London School of Hygiene and Tropical Medicine, Keppel Street, London WC1E 7HT*). Dr Dresch has rightly emphasized the importance of a multi-disciplinary approach to the problems of the Sahel semi-arid region; so many of the factors that have interacted in contributing to the drought appear to have little obvious scientific relation to one another.

As he has stated, famines have occurred in the area before: they are recorded by the Tuareg names such as 'Forget your wife' or 'The sale of children', indicating the terrible privations that they underwent; but previous droughts were limited to certain regions and none showed rainfall below average for longer than three years. The recent drought covered much of the Sahara, the whole Sahelian zone and much of the Sudan zone, and continued with below average rainfall for at least six years. This drought seems to have been in a different category from previously recorded droughts.

I think that Dr Dresch perhaps underestimates the economic importance of Sahelian cattle grazing. The Sahel is substantially disease-free, and the Fulani arrange their migrations so that calves are dropped there during the summer rains. Very large populations of their Zebu cattle have been built up which supply the main needs of the coastal towns. The cattle trade has increased the economic prosperity of the region and increased the demand for agricultural production, which may have played an important part in the over-grazing and deterioration of the environment. Unfortunately many of the methods that Dr Dresch has proposed as a remedy, such as wells and increase in agriculture, have already tended to exacerbate the situation.

Phil. Trans. R. Soc. Lond. B. **278**, 543–554 (1977) [543]
Printed in Great Britain

Evaluation of land-use and land-treatment practices in semi-arid western United States

By R. F. Hadley
U.S. Geological Survey, Denver, Colorado, U.S.A.

There are about 81000000 hm^2 of arid to semi-arid rangeland in the 11 Western States of the United States. Much of this area was subjected to poor land-use practice in the early days of settlement resulting in deterioration of plant cover and severe erosion. Beginning in the 1930s land-treatment programmes were initiated to restore the range and control runoff and erosion. Results of investigations designed to evaluate the effects of land-use and land-treatment programmes are discussed. The types of practices considered are: (1) grazing control, (2) vegetation modification, (3) mechanical land treatments and (4) water spreaders. Grazing control alone reduced runoff by 30 % and sediment yield by 35 % in the Badger Wash basin of western Colorado. Vegetation conversion from shrubs to grass generally does not increase water yield, however, sediment yields are reduced as much as 14 times as shown by a controlled experiment at Boco Mountain in central Colorado. Of the many types of mechanical land treatments designed primarily to increase infiltration, the most effective is contour furrowing. Water spreading on valley floors by diverting flood flows increases forage production, reduces flood peaks and sediment loads to downstream areas.

All treatment practices designed to improve plant cover, induce infiltration, or control flood flows will use water. Therefore, these practices must be evaluated with regard to their effect on water yield, which is often critical in semi-arid regions.

Introduction

Much of the land area in the 11 Western States, outside the high mountain forests, is in a climatic environment that is arid or semi-arid. Although these lands generally can be classed as water deficient, they are of vital importance to the economy of the western United States. The total area of the 11 Western States is about 307572000 hm^2 (ha), of which over 81000000 hm^2 may be classified as arid or semi-arid rangeland. These rangelands, which are predominantly in federal ownership, are the subject of this paper. The dry climate combined with immature and commonly rocky soils and unfavourable topography precludes their use for agriculture. Most areas that are irrigable by reason of available water and favourable relief have already been developed for agriculture.

Most of these lands received 200–380 mm annual precipitation. Large areas in the arid regions of southern California, western Arizona, and southern Nevada, however, received less than 130 mm annually. The fact that much of the precipitation occurs as high-intensity summer thunderstorms coupled with high rates of evapotranspiration reduces further the available soil moisture for plant growth.

The same factors which in the aggregate make the land fit only for grazing also tend to make it vulnerable to erosion. The low annual precipitation supports a sparse plant cover that is not very effective as a protection against erosion. Periods of drought that take their toll of the fragile plant cover are often interrupted by high-intensity rains that produce destructive flooding. All of these factors make it imperative that management and use of land resources in arid and semi-arid environments be based on an understanding of hydrologic and geomorphic processes and soil–plant–water relationships.

History of land use

In addition to the physical, hydrological, and climatological characteristics that make the arid and semi-arid rangelands highly susceptible to erosion, there has been a long history of land abuse by overgrazing. From the middle part of the 19th century until 1934, all public domain lands in the West were open to use by anyone without restriction or regulation. The result was indiscriminate use by livestock and wildlife during all seasons of the year without consideration of grazing capacity or maintenance of the plant cover. By the 1870s, the livestock populations had reached, and passed, the safe grazing capacity of the range. This period was climaxed by a critical drought and the severe winters of 1885–6 and 1886–7 (U.S. Dept. of the Interior 1975). Livestock losses were so severe that some large ranching companies failed. In this same period, large areas of public land on the Great Plains, which at best were marginally suitable for agriculture, were being opened to homesteading. When farming proved unsuccessful, the virgin prairie now broken by the plough only added to the erosion and rangeland depletion problems initiated by overgrazing in the previous decades. The range condition continued to deteriorate and probably reached a low point in the 1920s (Clawson 1972).

The Taylor Grazing Act, passed by the U.S. Congress in 1934, provided the necessary legislation to regulate the use of public lands for grazing. The passage of the Taylor Act coincided with the severe drought then prevailing throughout much of the West that affected rangeland as well as farmland. In 1932, it was estimated that the western range had lost nearly 50% of its original productivity (U.S. Dept. of the Interior 1975). Because of the drought, attention was focused on erosion and soil conservation problems on the deteriorated semi-arid rangelands. In the period since about 1936, many programmes have been initiated to improve rangelands and control erosion and sedimentation. Many of these programmes were not successful because they were started without adequate hydrologic data or an understanding of hydrologic processes in the semi-arid environment (Peterson & Hadley 1960; King 1968). In recent years, however, controlled watershed experiments have made it possible to evaluate a variety of land-use and land-treatment practices representing a wide range of natural conditions of soil, vegetation, and topography.

Trends in range use and condition

If land-treatment practices are to be successful, the control of grazing is essential. With the establishment of grazing districts in 1936 and the assignment of individual grazing allotments to range users, itinerant herds were eliminated and major changes in grazing use began. Data on grazing before 1940 are almost non-existent but it is known that significant reductions in numbers of grazing animals occurred following passage of the Taylor Act 1934. For example, Peterson & Hadley (1960) state that between 1940 and 1955 in the upper Colorado River basin there was a 25% reduction in the animal-unit months of permitted grazing on public lands. These reductions were made on a voluntary basis, marking the beginning of a general awareness that the range had been overstocked and could not continue to be used so heavily. Reductions in numbers of animal-unit months permitted on grazing district lands has continued since 1955 as shown in table 1. Accompanying the more judicious use of the public lands for grazing has been an improvement in forage condition and soil erosion condition. Although much of the information on trends is qualitative, it probably reflects the effects of land management in a general way. Tables 2 and 3 summarize information published by Deming (1960) and the U.S. Dept. of the Interior (1975).

TABLE 1. PERMITTED USE OF GRAZING DISTRICT LANDS, 1960–1970†

State	1960	1964	1970
		Animal units‡	
Arizona	126568	105026	95050
California	97154	90704	78442
Colorado	296707	335466	332564
Idaho	393031	388525	399945
Montana	475116	440814	349807
Nevada	425057	451708	432224
New Mexico	347989	356682	285347
Oregon	264764	238778	201279
Utah	368501	344658	299394
Wyoming	416122	443711	437701
	3211009	3196072	2911753

† Adapted from Public Land Statistics.
‡ One cow or one horse or five sheep or five goats, all over 6 months of age.

TABLE 2. TRENDS IN RANGELAND CONDITION, PUBLIC DOMAIN IN WESTERN U.S.

forage condition	1930–35 ha	1930–35 per cent of total	1955–59† ha	1955–59† per cent of total	1975 ha	1975 per cent of total
improving	507898	1	14771550	24	12550961	19
unchanged	3182561	6	35047020	57	42873513	65
declining	48026963	93	11638363	19	10418597	16
total	51717422		61456933		65843071	

† Adapted from Deming (1960).

TABLE 3. SOIL EROSION CONDITION CLASSIFICATION FOR 58479150 ha OF PUBLIC LANDS IN WESTERN U.S., 1975†

	condition class				
	stable	slight	moderate	critical	severe
hectares	5263123	26900409	20467702	5263123	584791
per cent	9	46	35	9	1

† Bureau of Land Management (1975).

All conservation measures designed to control erosion or improve plant cover will use water by detention in reservoirs, mechanical improvement of infiltration, control of overland flow on upland slopes by mechanical treatments, or diversion of flood flows in channels for irrigation of bottomlands. Therefore, evaluation of these practices must include an assessment of on-site benefits as well as the effects on the water users downstream. Competition for water in semi-arid regions is constantly increasing for a variety of uses such as urban development, mineral development, recreation, and wildlife management. It is important to note, however that many of the areas with serious erosion problems generally have a meagre water yield when compared with the annual discharge of the entire river basin whereas sediment loads are often extremely high. If erosion can be effectively controlled in upland areas by a combination of grazing practices and mechanical or structural treatments, the rangelands may be measurably improved, resulting in reduced sediment loads. Results of studies considered here, however, indicate that this cannot be done without a consumptive use of water.

Effects of grazing

In order to verify the hypothesis that overgrazing or unregulated grazing on semi-arid rangelands has increased erosion, sediment yields, and destructive flooding, studies have been made on the hydrologic effects of grazing practices (Lusby, Turner, Thompson & Reid 1963; Lusby, Reid & Knipe 1971; Dunford 1949). Of the treatment practices that can be initiated to improve range condition and control runoff and sediment yield, grazing management is the simplest to apply to entire grazing units, but perhaps the most difficult to evaluate without strictly controlled experiments.

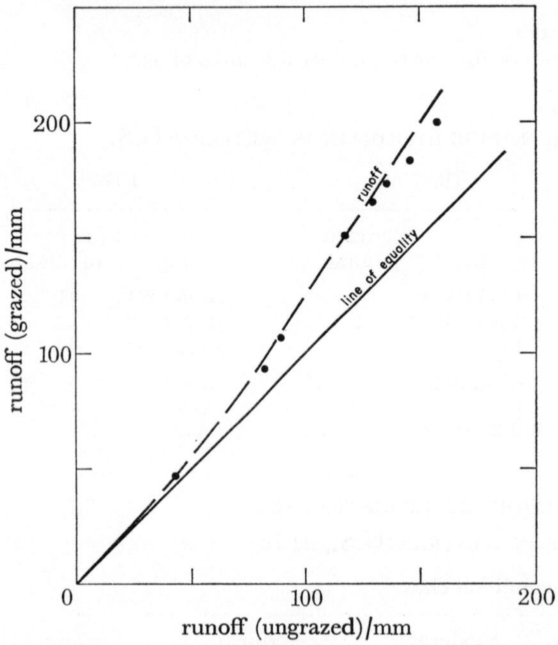

Figure 1

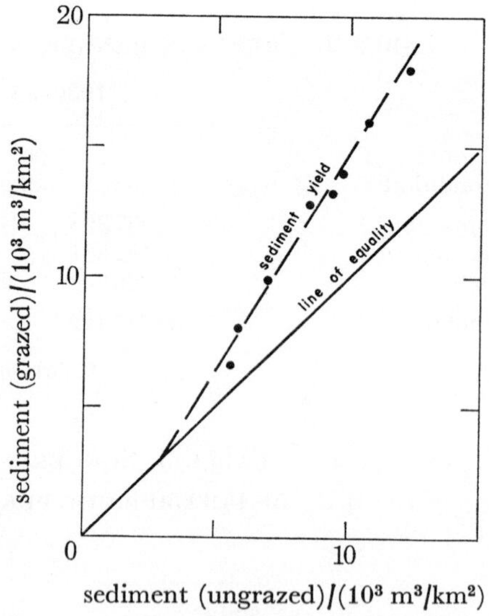

Figure 2

Figure 1. Mass diagram of runoff from four grazed and four ungrazed drainage basins in Western Colorado. The data represent 13 years of records. (Adapted from Lusby *et al.* 1971.)

Figure 2. Mass diagram of sediment yields from four grazed and four ungrazed drainage basins in Western Colorado. The data represent 13 years of records. (Adapted from Lusby *et al.* 1971.)

A long-term study of the effects of grazing on hydrology was conducted in Badger Wash basin in western Colorado from 1954 to 1973. Four pairs of small drainage basins were studied; one basin of each pair was grazed and one basin was fenced and excluded from grazing. The treatment was unchanged during the first 13 years of the study, 1954–66, and the results as described by Lusby *et al.* (1971) are summarized here.

The climate of the Badger Wash basin is arid to semi-arid with an average annual precipitation of 220 mm. Vegetation is of the salt-desert shrub type and the cover is sparse on soils that are predominantly fine-grained. In the period 1954–66, runoff from three grazed basins averaged 131–140 % of that from three ungrazed basins (figure 1) and during the last 6 years of the period, runoff from grazed basins averaged 140–145 % of that from ungrazed basins. The greatest change occurred in the runoff relation after the third year of grazing exclusion. Similarly, the sediment yield from grazed basins was 50 % higher than from ungrazed basins (figure 2) (Lusby *et al.* 1971). The major factor influencing the marked differences in run-off and sediment yield appears to be the compaction of soils by grazing animals in the spring season when the surface materials are damp and loose from winter frost action. This causes increased runoff and erosion on grazed basins. This observation was verified in the period 1966–73 when the grazing season was changed, which had essentially the same effect on runoff as eliminating grazing completely (Lusby 1976, personal communication). A similar relation between grazing intensity and runoff was reported on plots in east-central Colorado by Dunford (1949). These studies indicate that grazing management alone can accomplish control of runoff and sediment yield.

Vegetation modification

As already stated, the plant cover was seriously deteriorated on much of the semi-arid rangeland of the western United States in the 1930s by a combination of factors attributable to overgrazing and drought. Much of the mixed sagebrush–perennial grass rangeland soon became dominated by sagebrush with very little grass understory as the competition for soil moisture was accentuated by the drought. Many conservationists felt that vegetation modification by conversion from shrubs or trees to grassland would not only improve the forage production on the range but would reduce erosion and sediment yields and possibly increase water yield. Many of the early predictions of what could be beneficially accomplished by vegetation conversion were intuitive rather than being based on controlled experiments. In recent years, several studies have been made that quantitatively evaluate the environmental effects of vegetation conversions from woody to herbaceous species. A discussion of the advantages of shrubs in terms of palatability for livestock and wildlife and an analysis of the indictment that shrubs are worthless invaders is beyond the scope of this paper. For an excellent treatment of this subject the reader is referred to a paper by McKell (1975) regarding the usefulness of shrubs as a resource. Some of the effects of vegetation conversions on runoff and sediment yield are considered here.

Several million hectares of sagebrush have been converted to grassland in the Western United States in the past 25 years. The most common methods of removing the shrubs are ploughing or spraying with chemical herbicides. Following the elimination or control of the sagebrush, the rangeland is seeded to perennial grasses with the objective of increasing forage production and controlling erosion. Results of a study of sagebrush conversion to grassland by Shown, Miller & Branson (1969) show that the most successful conversions usually occur where the annual precipitation exceeds 254 mm and on soils having medium moisture-holding capacities such as sandy loams, loams, or silt loams.

In a controlled watershed experiment conducted by the Geological Survey, the effects of conversion from big sagebrush cover to wheatgrass cover on runoff and sediment yield were investigated during a 9-year period, 1965–73. Four small basins, 2.02 to 4.04 ha, located about 4.8 km north of Wolcott in western Colorado were used for the study. The study area is named Boco Mountain.

Annual precipitation at the study area is about 343 mm; about 229 mm occur as rain during April through October, and about 114 mm occur as snow that accumulates as a snowpack during November through March (Shown 1971). Runoff and sediment yield from four study basins were measured in reservoirs to evaluate the effects of the treatment. After a calibration period of 3 years, the sagebrush was removed from two of the basins by disk-ploughing and those basins were seeded to wheatgrass (Shown 1971).

During the calibration period, 1965–7, the average annual runoff from the four study basins was 33.8 mm and the average annual sediment yield was 962 m^3/km^2. For the 6-year period, 1968–73, after two basins had been converted from sagebrush to wheatgrass the average annual runoff for the two sagebrush basins was 29 mm and for the two grassed basins it was 32 mm. Therefore, the results indicate that the vegetation conversion did not affect the total runoff from rainfall and snowmelt appreciably (G. C. Lusby 1976, personal communication). However, there was a considerable reduction in sediment yield. The two basins left in sagebrush cover had an average annual sediment yield of 266 m^3/km^2 during the 6-year post-treatment period while the two grassed basins yielded only 38.1 m^3/km^2 which represents a large decrease in sediment yield on the treated basins over that measured during the calibration period.

There have been few studies of the hydrologic effects of conversion of shrubs to grass such as the Boco Mountain study. However, observations in semi-arid regions indicate that such vegetation manipulation will not result in large increases in water yield. Also, more study is needed regarding the long-term effects of vegetation conversion on runoff, sediment yield, and forage production.

Mechanical land treatments

Various mechanical treatments have been used on rangelands in semi-arid regions of the Western United States. The most common of these practices is contour furrowing with the principal objective of increasing infiltration to enhance soil-moisture storage and consequent forage production on upland slopes. In an evaluation study of mechanical land treatments in the Western United States by Branson, Miller & McQueen (1966), they conclude that contour furrowing on medium- to fine-textured soils is the most effective type of treatment for conserving soil moisture. According to Caird & McCorkle (1946), 404 700 hm^2 of pasture and rangeland were contour furrowed between 1936 and 1940. The practice has been continued since 1940, and the area treated has perhaps doubled. As with structural treatments in stream channels and on valley floors, mechanical treatments on upland areas tend to impede overland flow and reduce flood peaks, thereby increasing on-site consumptive water use.

Benefits derived from contour furrowing are documented by the studies of Branson *et al.* (1966). They state that forage yields were increased by 118% in treated areas in Montana and by 136% in treated areas in Wyoming. When newly constructed, the water storage capacity of furrows generally exceeds 51 mm of precipitation. Assuming infiltration rates of at least 13 mm/h, the storage capacity should not be exceeded in semi-arid regions where a 2-h rainfall of 51 mm has an expected recurrence interval of 50 years.

As emphasized by Branson *et al.* (1966) the storage capacity of contour furrows decreases with time. Figure 3 shows that the decrease in storage capacity tends to be rapid during the first 5 years and becomes relatively stable after about 9 years. However, when the plant cover becomes well established, the effectiveness of the treatment as an erosion and runoff control measure apparently lasts for more than 20 years.

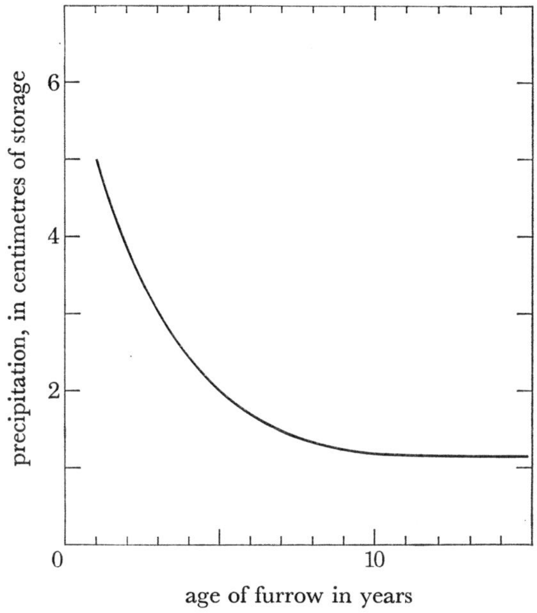

FIGURE 3. Decrease in water storage of contour furrows with age. Storage capacities were computed from measurements of depth, width, and spacing of furrows. (Adapted from Branson *et al.* 1966.)

WATER SPREADERS

Water spreading on alluvial valley floors adjacent to ephemeral stream channels is a common practice in semi-arid regions to increase forage production, reduce flood peaks, and induce aggradation in trenched valleys. In a study by Miller *et al.* (1969) of water spreaders in the western United States it was found that the practice was most successful in areas receiving 280 mm or more of precipitation annually and was not successful in areas receiving less than 203 mm of precipitation annually.

The primary objective of a floodwater spreader is to divert flood flows from a trenched ephemeral-stream channel onto adjacent flood plains and low terraces. Many of these valley floors were flooded naturally in the past before the alluvium was trenched by erosion. Depletion of the vegetation by overgrazing caused the alluvial deposits to become more susceptible to erosion. Deepening and headward cutting of the gully trench not only lessened the frequency of overbank flooding but also decreased the soil moisture available for plant growth because the gullies served as drainage ditches. In order to restore the irrigation of the bottomlands, which are the most productive lands in the semi-arid regions, ranches soon began devising ways to utilize the flood flows by diverting water onto the valley floors.

There are two common methods of water spreading: (1) direct diversion of flood flow from the channel by constructing earthfill dams across the trench and allowing natural flooding of the valley floor, and (2) construction of a detention dam and reservoir with a series of training dykes that distribute the water back and forth evenly across the valley floor. In both designs the important consideration is to get the water on and off the spreading area without ponding for long periods. Ponding will cause deposition of very fine sediment and prolonged inundation that will kill the existing plant cover.

The operation of a water spreader was evaluated in a study by Hadley & McQueen (1961) in the Box Creek basin of east-central Wyoming in an area that receives about 356 mm annual precipitation. The Box Creek water spreader is operated by direct diversion of ephemeral flows from the main channel by a series of 27 earthfill dams. Natural flooding of an area of about 146 ha on the valley floor occurs along a reach of the stream channel about 6.4 km long. The drainage area upstream from the water spreader is about 363 km^2.

Gauging stations were established upstream and downstream from the water spreader and suspended-sediment samples were taken at both gauging stations during the 1956 and 1957 seasons (April through September) for which inflow and outflow through the water-spreading system could be determined. The total inflow for six runoff events was 2.50 hm^3 and the outflow was 1.64 hm^3, which represents a loss of 34 % of surface flow entering the water spreader.

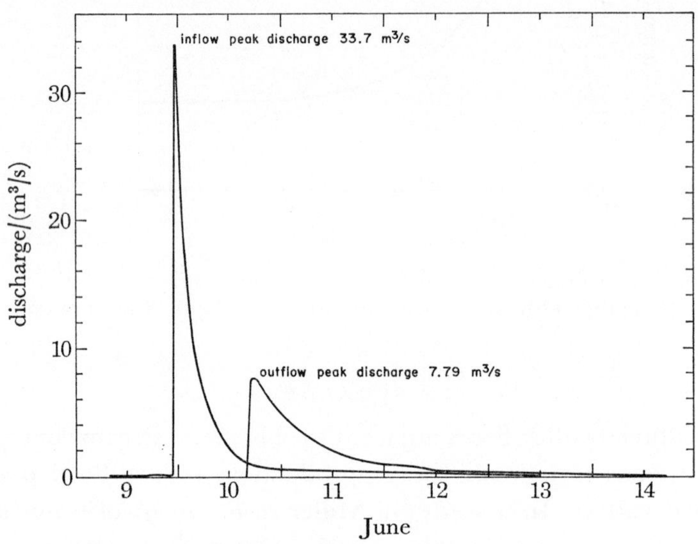

FIGURE 4. Flood hydrographs for gauging stations at Box Creek water spreader for the period 9–14 June 1957.

There were two storms during the study period that centred in the headwater reaches of Box Creek, and all flow in the channel originated above the water-spreading system. These runoff events occurred 9–14 June and 30 June–4 July 1957. Data collected on streamflow and suspended-sediment discharge during these two periods provide a good example of the operation of the water spreader. Inflow to the water spreader during the 9–14 June period was 0.626 hm^3 and the instantaneous peak discharge was 33.7 m^3/s. Outflow from the water spreader was 0.434 hm^3 and the instantaneous peak discharge was only 7.8 m^3/s. The inflow and outflow hydrographs for the 9–14 June runoff event are shown in figure 4. During the runoff event of 30 June–4 July 1957, the inflow to the water spreader was 0.59 hm^3 and outflow was 0.44 hm^3. The instantaneous peak discharge at the upper gauging station was 23.8 m^3/s and the outflow instantaneous peak discharge was only 7.0 m^3/s.

During the 9–14 June period, the suspended-sediment load entering the water spreader was 2460 t and at the outflow station the load was only 752 t. In the period 30 June–4 July, the suspended-sediment load at the inflow station was 1633 t and at the outflow station was only 263 t. Therefore, during the two runoff events there was a depletion of suspended-sediment load through the water spreader of 3078 t, or 75 %, by deposition in the channel and on the valley floor. The histograms of suspended-sediment load shown in figure 5 illustrate graphically the effect of the water spreader on suspended-sediment discharge.

As has been shown in other studies in semi-arid regions (Culler 1961; Hadley & Schumm 1961), generally there is a natural depletion of stream flow and sediment during runoff periods in ephemeral streams. Operation of water-spreading systems, such as the Box Creek water spreader, will reduce peak discharges and increase the expected flow losses and sediment

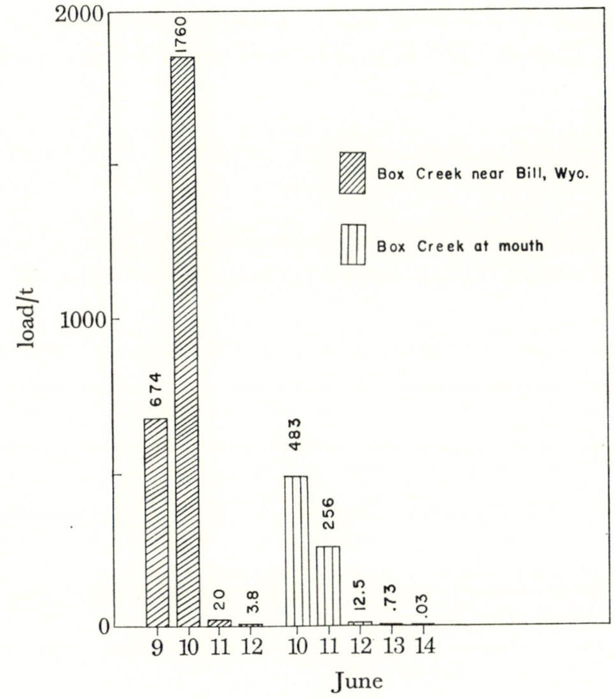

FIGURE 5. Histograms of suspended-sediment load at gauging stations on Box Creek for period 9–14 June 1957.

deposition. Also, it should be noted that the Box Creek water spreader was 30 years old at the time of the study by Hadley & McQueen (1961). The trenched channel had been nearly filled with sediment and the plant cover on the valley floor was well recovered. New systems in valleys with deep trenches, reduced soil moisture, and poor plant cover may be expected to deplete flood flows even more than the observed measurements in Box Creek.

LAND USE AND HYDROLOGY

Many of the land-use and land-treatment practices that have been described are successful in the control of runoff, abatement of erosion, and increase of plant cover but generally on relatively small and widely scattered upland areas. When compared with the millions of hectares that are in need of improved land-use practices, it seems improbable that the impact of these conservation measures would be detectable in the hydrologic response of the major streams. However, there are many examples in the literature (Hadley & Schumm 1961; King 1968; Emmett 1974; Hadley 1974) of recent changes in channel morphology and sediment discharge of streams in semi-arid regions of the western United States. Although our information on land-treatment practices and changes in range condition since 1930 are far from adequate, the evidence suggests that reductions in sediment yields are a reflection, at least in part, of improved land use.

In a study by Hadley & Schumm (1961) in the Cheyenne River basin, eastern Wyoming, there are numerous examples of channel and flood plain aggradation. On Twentymile Creek, near Lusk, Wyoming, the extent of overbank deposition was determined by measurement of buried fence posts along a line that crosses the valley floor. The flood plain had been aggraded to a depth of about 1 m throughout a reach that is 3.2 km in length and approximately 460 m wide in a period of 31 years, 1920–51. In the same period about 100 small reservoirs were built in upland tributaries of Twentymile Creek for the purposes of providing livestock water and erosion control. These structures undoubtedly have reduced flood peaks and flood volumes and induced deposition of sediment on the valley floor.

TABLE 4. AVERAGE ANNUAL DISCHARGE AND SUSPENDED-SEDIMENT LOAD AT GAUGING STATIONS IN THE UPPER COLORADO RIVER BASIN FOR TWO PERIODS OF RECORD (after Hadley 1974)

gauging station	period† 1	discharge hm³	suspended sediment load 10^3 t	period† 2	discharge hm³	suspended sediment load 10^3 t	ratio period 2/period 1: discharge	ratio period 2/period 1: sediment
Colorado River at Grand Canyon, Arizona	1926–41	15460	177300	1942–60	14500	86400	0.944	0.486
Colorado River near Cisco, Utah	1930–42	6340	17300	1943–62	6170	8700	0.973	0.501
Green River at Green River, Utah	1930–42	4490	22100	1943–62	5200	15100	1.156	0.683
San Juan River near Bluff, Utah	1930–42	2430	41700	1943–62	1880	15900	0.775	0.381

† Water year (1 October–30 September).

In the Wind River basin of Central Wyoming, King (1968) reports a similar example of induced aggradation by control of runoff in two adjacent drainage basins, Logan Draw and Conant Creek. Logan Draw has a drainage area of 157 km² and Conant Creek has a drainage area of 153 km² upstream from the junction with Logan Draw. In 1953, the Bureau of Land Management completed a flood control programme on Logan Draw consisting of diversion dams and water spreaders that completely utilized all flood flow from the basin.

Since 1953 no flow has passed the mouth of Logan Draw into Conant Creek. The result has been that the channel of Conant Creek, which was adjusted to floods from both basins, began to aggrade. Surveys of channel cross sections by King during the period 1955–66 show that the channel filled to a depth of 1.2 m and total deposition was about 10200 m³ in a reach of Conant Creek 1 km in length.

These examples of channel aggradation are not unique. Emmett (1974) and Hadley (1974) report on observations at several locations in the Missouri, Rio Grande, and Colorado River Basins of aggrading channel reaches in the past 20–30 years.

The observations of improved plant cover and sediment deposition in smaller tributaries of the western United States river basins is also manifested regionally. For example, comparison of the suspended-sediment discharge records at the Grand Canyon gauging station on the

Colorado River for two periods, 1926–41 and 1942–60, show that the suspended-sediment loads for the latter period are only about 50 % of the earlier period (Hadley 1974). A similar relationship is found in the records of two major tributaries of the Colorado River, Green River and San Juan River. Although the periods of record are not exactly the same there is sufficient overlap for comparison. The records are summarized in table 4 (Hadley 1974).

The reduction in sediment yield in the Upper Colorado River basin between 1942 and 1962 coincides with the period when land-use and land-treatment practices were initiated and grazing regulations began on public lands. Admittedly, it is difficult to determine the effects of individual treatment practices, whether they consist of structural control, mechanical treatment, or grazing plans. Nevertheless, it is evident that progress has been made in the management of the limited land and water resources of semi-arid rangelands in the western United States.

References (Hadley)

Branson, F. A., Miller, R. F. & McQueen, I. S. 1966 Contour furrowing, pitting, and ripping on rangelands of the Western United States. *J. Range Mgmt* **19** (4), 182–190.

Caird, R. W. & McCorkle, J. S. 1946 Contour-furrow studies near Amarillo, Texas. *J. Forestry* **44**, 587–592.

Clawson, M. 1972 *America's land and its uses.* Baltimore: Johns Hopkins Press.

Culler, R. C. 1961 Hydrology of stock-water reservoirs in upper Cheyenne River basin. *U.S. Geol. Survey Water-Supply Paper*, 1531-A, pp. 1–136.

Deming, M. H. 1960 Conditions and trends surveys. *Our public lands.* U.S. Bureau of Land Management, October.

Dunford, E. G. 1949 Relation of grazing to runoff and erosion on bunchgrass ranges. *Res. Note, Rocky Mountain Forest & Range Experiment Station* **7**.

Emmett, W. W. 1974 Channel aggradation in Western United States as indicated by observations at Vigil Network sites. *Z. Geomorph.* (Suppl.) **21**, 52–62.

Hadley, R. F. 1974 Sediment yield and land use in Southwest United States. *Int. Ass. Hydrol. Sci. Pub.* no. **113**, pp. 96–98.

Hadley, R. F. & McQueen, I. S. 1961 Hydrologic effects of water spreading in Box Creek basin, Wyoming. *U.S. Geol. Survey Water-Supply Paper*, 1532-A-22-A-35.

Hadley, R. F. & Schumm, S. A. 1961 Sediment sources and drainage basin characteristics in upper Cheyenne River basin. *U.S. Geol. Survey Water-Supply Paper*, 1531-B, 137–197.

King, N. J. 1968 Restoration of gullied valley floors in arid and semiarid regions. *23rd Int. Geol. Cong.* **12**, 187–195.

Lusby, G. C., Reid, V. H. & Knipe, O. D. 1971 Effects of grazing on hydrology and biology, Badger Wash basin, Colo., 1953–66. *U.S. Geol. Survey Water-Supply Paper*, 1532-D, pp. 1–90.

Lusby, G. C., Turner, G. T., Thompson, J. R. & Reid, V. H. 1963 Hydrologic and biotic characteristics, Badger Wash basin, Colo., 1953–58. *U.S. Geol. Survey Water-Supply Paper* 1532-B, 1–73.

McKell, C. M. 1975 Shrubs – a neglected resource of arid lands. *Science, N.Y.* **187**, 803–809.

Miller, R. F., McQueen, I. S., Branson, F. A., Shown, L. M. & Buller, W. 1969 An evaluation of range floodwater spreaders. *J. Range Mgmt* **22** (4), 246–257.

Peterson, H. V. & Hadley, R. F. 1960 Effectiveness of erosion abatement practices on semiarid rangelands in Western United States. *Pub. Int. Ass. Sci. Hydrol.* **53**, 182–191.

Shown, L. M. 1971 Sediment yield as related to vegetation on semiarid watersheds. *Proc. 3rd Int. Seminar for Hydrology Professors*, 347–354.

Shown, L. M., Miller, R. F. & Branson, F. A. 1969 Sagebrush conversion to grassland as affected by precipitation, soil, and cultural practices. *J. Range Mgmt* **22** (5), 303–311.

U.S. Dept. of the Interior 1970 *Public land statistics.* Washington: U.S. Government Printing Office.

U.S. Dept. of the Interior 1975 Range condition report prepared by the U.S. Bureau of Land Management, 2-7-2-14.

Discussion

B. Dasgupta (*Institute of Development Studies, University of Sussex*). As the speaker is surely aware, while many of the measures suggested by him are technologically sound, these may not be feasible in financial or economic terms. Would the speaker be kind enough to indicate, approximately, the total cost incidence of various schemes and how the money was raised?

N. E. Reynolds (*The Ford Foundation, 55 Lodi Estate, New Delhi* 110 003, *India*). Dr Gupta's statement that 'third world countries like India cannot afford conservation works' is largely unexplained and cannot be accepted. In India large sums are spent on relief and this money can pay for works whose calculated rates of return may be low and otherwise unacceptable. The proper question is the administration of relief and, to go further, of general local administration. Nor should conservation be treated in isolation. In India the large areas of forest and grazing land destroyed over the last 50–60 years testify to the importance of legal and institutional factors, factors that remain central when considering conservation and land-use issues in populated areas. Another set of factors is: employment as a basic right (Maharashtra's Guarantee Employment Scheme); as an economic development strategy; and as a means to the husbandry and development of scarce resources.

Conservation work is surrounded by difficulties in poor countries. Contour lines do not fit existing field and farm boundaries so that, without land consolidation, work on the contour is not easy. The power structure within watersheds tends to reside at the bottom, amidst the better soils, more level lands, stream and communication lines. Erosion of upland in the early stages is not seen as a threat to the powerful. Siltation, indebtedness and greater instability can be to the advantage of lower level farms if those farmers can secure the services of the conservation department to build siltation works, and those farmers may be able to command a more docile labour supply by creating greater dependency through money-lending, etc. Good conservation is often a good distributive measure. To do it in that manner requires a strong political leadership supported by confident and determined disciplinary groups and departments.

S. Sandford (*Overseas Development Institute*, 10–11 *Percy Street, London W*1). In evaluating the success of land-treatment practices, Dr Hadley had quoted figures for reductions in runoff and in sediment-load as his criteria. Do figures also exist to show the effect on economic output, possibly measured in terms of quantity of meat?

R. F. Hadley. Beneficial effects of contour furrowing, rangeland pitting, and other mechanical land treatments that conserve soil moisture are all generally similar. In addition to reducing runoff and sediment yields they increase forage production and livestock growth. It has been reported in a study in southeastern Wyoming that effects of rangeland pitting increased sheep carrying capacity by about 32 % and lamb gains by about 11 kg/ha per year.

E. B. Worthington (*I.B.P., c/o The Linnean Society, Burlington House, Piccadilly, London W*1*V* 0*LQ*). What is the effect of 'water spreading' to encourage the growth of fodder plants on the quality of the runoff water downstream, or, in cases where all the spread water is transpired locally, on the soils? Would not the salt or alkaline content accumulate with time in a way comparable to older irrigation schemes in arid and semi-arid areas?

R. F. Hadley. Water quality is generally improved downstream from a water spreader. Sediment discharge is decreased by reduced velocities and deposition in the spreader system. Because of annual deposition of fresh sediment and leaching of salts in the medium-textured soils of successful water spreaders salinity problems are minimal. In poorly-designed water spreaders where water is ponded for long periods and soils are very fine grained, however, salinity problems do exist.

Phil. Trans. R. Soc. Lond. B. **278**, 555–563 (1977) [555]
Printed in Great Britain

Land-use in semi-arid southern Africa

By H. C. Pereira, F.R.S.

The Author was formerly Director of the Agricultural Research Council of Central Africa, and earlier Deputy Director of the East African Agriculture and Forestry Research Organization. He is now Chief Scientist, Ministry of Agriculture Fisheries and Food, London

[Plate 1]

Africa south of the Equator has a low population density which reflects the scarcity of reliable supplies of potable water. Productive agricultural methods have been demonstrated and are practised on advanced enterprises throughout the sub-continent, but most of the population still live by subsistence agriculture which is increasingly destructive of soil and water resources as numbers increase. Higher priority for agriculture, better rewards and status for food producers and the urgent application of known techniques of soil conservation are all necessary if the semi-arid areas of southern Africa are to feed their increasing populations.

Southern Africa is an instructive sub-continent in which to study land use in semi-arid climates, since it provides current examples of the widest possible range of practices, from the subsistence patterns of pre-history to the most sophisticated large-scale irrigation developments. In remote areas of northern Zambia, tribal patterns of shifting cultivation in the open Brachystegia woodland on geologically old leached soils, still depend on the lopping of tree branches over a wide area, burning them in a central clearing, and growing two or three crops of millet on the areas temporarily enriched by the ash. It is believed that this practice has continued unaltered from the distant past (Trapnell 1953; Allen 1965; Lawton 1963). In contrast, the scientific development of the Vaal and Orange river system with its many dams, diversion weirs, canals, tunnels and pipe-lines, ranks with the Tennessee Valley in the intensity of combined hydro-power and irrigation development.

The geography of the sub-continent is also full of contrasts, from high altitude mountain forests to hot coastal deserts. I must remind you that most of the area is a high plateau, 1000 m or more above sea level, so that temperatures are a little modulated by altitude, and the hottest semi-arid areas are near the lowland coasts of Kenya and Somaliland and of SW Africa on the edge of the Kalahari desert.

For the purposes of this brief paper I shall take the Equator as my northern limit, although I shall quote some examples from the semi-arid areas of Ethiopia a little to the north.

The climax vegetation, to which more than half of the area reverts when undisturbed, is of open woodland, or tropical grasslands carrying varying densities of scattered trees, and known collectively as savanna. It has been ecologically described and classified but in this short talk I must concentrate on how it is used.

The human population looks sparse on a world map of population density, and indeed it averages less than 4 per square mile (1.5/km^2) overall but, as a world map of poverty shows, Africa represents the largest geographical area of incomes less than $300 p.a. The total numbers are small, and yet their uneven distribution overcrowds favourable areas to the point of soil destruction, while large areas are, effectively, uninhabited.

This distribution of human settlement is determined mainly by water supplies, both as rainfall to sustain vegetation and as surface water or potable groundwater for people and livestock. The characteristically high evaporation rates of 1.5–2 m per annum, combined with the erratic annual movement of the inter-tropical convergence zone which causes the main seasonal rains, together render the reliability of the rainfall more important than the long-term annual average. For example, Glover, Robinson & Henderson (1954) showed that the ecological boundaries of vegetation types in East Africa fitted far more closely to the areas in which rainfall of a given amount was received in seven years out of ten than to the mapped areas of average annual rainfall. The boundaries of reliance on 750 mm outlined the maize-growing areas, while in areas for which 500 mm could be expected, sorghum is the main crop. An intriguing result of the mapping of Tanzania is that the corridor of reliable rainfall outlines the historical slave-trade route from Zanzibar to Lake Tanganyika, which evidently followed the course of the most reliable food and water supplies.

Rain seasons of erratic duration and amount are followed by severe dry seasons in which desiccation of the vegetation and of exposed soil leaves the surface vulnerable to soil erosion. Indigenous trees, shrubs and grasses root deeply and widely. When assessing semi-arid land for productivity, the soil depth and capacity to store water sufficient to mature a crop or to maintain grass growth for grazing is of far more critical importance than are the soil nutrient reserves. The nutrients can be reinforced by fertilizer, while resources for irrigation are available only for a small proportion of the land.

This liability to a premature cessation of the rains and hence to a shortening of the growing season indicates two major thrusts of agricultural research and development. The first is to select for rapid growth of crops and livestock, and the second is to develop supplementary irrigation wherever possible.

The interaction of energy reflected from a hot land surface with the wet oceanic air of the convergence zone results in characteristically violent storms with high rainfall intensities. These effects are more acute as the rainfall total decreases, while the vegetation protecting the soil is more sparse. Thus, paradoxically, the less the amount of annual rainfall the greater the erosion damage tends to become.

Potential productivity

Ample available solar energy and freedom from frost give almost the whole of southern Africa a basic advantage over temperature-zone farmlands. That we already have the knowledge and skills to develop these advantages into highly productive agriculture is widely demonstrated throughout the sub-continent. Scattered throughout all these countries, from the Equator to the Cape, are large estates with efficient science-based management and high agricultural output. Some indeed have high capital expenditure on irrigation but many more have grown from small beginnings with a main characteristic of managerial skill and resourceful application of scientific advances in tropical agronomy. I am not referring to the plantation crops of tea, coffee and oilpalm, which are confined to the higher and wetter country, but to maize and livestock enterprises, to sisal in the 500 mm rainfall, and to irrigated cotton, sugar and citrus. There is no doubt that the potential for agricultural production is high, but the fragile soil and vegetation cover need exacting standards of contour cultivation and soil conservation planning, of rotational grazing control and well distributed watering points.

The tribal traditions, however, have been founded on centuries of a scattered and rather stable population, nomadic herdsmen and shifting cultivators, with plenty of new land to use. There was no need to develop the elaborate terracing and water control as found in tropical areas of other continents. The tradition was to consume the accumulated natural fertility and to move on. Western medicine has now overcome the former population controls and Africa is now second only to Latin America in population growth; in fact, Swaziland, with 5.2 % p.a. tops the *U.N. Population Handbook* (1974). This population growth has already rendered the African tribal traditions of agriculture destructive to their environment, for there is now very little new land with water supplies, but still there is no general acceptance of the need to conserve the soil. A few heartening signs of response to agricultural teaching serve by contrast to emphasize the serious lack of progress in the tribal lands throughout the semi-arid areas of the sub-continent.

Technology for arable crops

The blunt statement that we already have enough proven technology to produce a vast increase in the productivity and prosperity of semi-arid areas of Africa may cause surprise, but there is abundant evidence for it. The main reason for this technical advance is that most of the same problems had already arisen acutely in the drier areas of the U.S.A. some 50 years earlier. In the U.S.A. agriculture is highly regarded as a source of wealth, and therefore the problems were tackled vigorously. The solutions were found to depend on complete watershed planning; on tillage and planting on the contour; on cut-off ditches to protect tilled soil from runoff from higher up the hillsides; on reducing the velocity of overland flow from heavy rainstorms by encouraging infiltration and by provision of grassed waterways, routed along natural drainage lines which freely cross farm boundaries.

When a majority of American farmers on a watershed voted for creation of an 'intensive conservation area' the rules became mandatory. These techniques were tested thoroughly in Kenya, which developed a vigorous soil conservation service on American lines, led by the redoubtable figure of Colin Maher of the British Colonial Service. In the areas of European settlement, the soil conservation rules were rigorously imposed. This service worked in the tribal areas also, but was less understood and less welcomed; conservation policies were pursued much more slowly, by persuasion only. Many square kilometres of the badly eroded Machakos area were thoroughly reconstructed from 1948 onwards with terraces and cut-off drains and for a while soil erosion was held at bay. Aerial photographs taken in 1948 have been carefully analysed by Thomas (1975) and compared with a later set taken in 1972. Thomas used the gulley-length measurement technique originated by Keech in Salisbury. The area had remained in subsistence agriculture in spite of study and demonstration that productivity could readily be doubled by simple improvements to management at a low level of inputs (Pereira & Thomas 1961). In spite of a terracing intensity of 75 m terrace length per hectare (330 yards per acre), cultivation onto steeper unprotected slopes had increased and the total area of severe erosion had increased from 26 to 37 % in the 24 years. The thin crops, grown without fertilizer or manure, gave so little protection that not even well-built terraces could succeed. Such fully documented examples are rare, but the general experience, uniform throughout Southern Africa, is that the whole standard of agricultural productivity must be increased simultaneously and that, although loss of soil is irrevocable and progressive, it is pointless to arrest it mechanically without tackling the root cause, which is the low agronomic standards of the arable cropping.

Irrigation techniques were also successfully adapted from overseas; the basic physics, which Howard Penman of this Society applied to the problems of Britain, were tested on a series of catchment area water balances in East Africa and shown to apply effectively in the tropics (Pereira 1959; Pereira *et al.* 1961). The essentials for irrigation management – evaporation rates as well as rainfall and temperature – have been studied, mapped and published (e.g. by Manning 1950; Griffiths 1960; Walker & Rijks 1967; Woodhead & Waweru 1970). Similar studies in the former Federation of Rhodesia and Nyasaland, continued in Rhodesia, Zambia and Malawi, and in South Africa (Bibliography by Rodier 1963). These make possible the more effective design of irrigation schemes, which have more probability of success today than many of the major irrigation schemes had 30 years ago in the U.S.A.

Major irrigation schemes have been developed successfully only where a fully commercial economy has been established; the largest schemes are the 1800 km Vaal River–Orange River System in South Africa, where there are many smaller schemes. The Sabi–Lundi scheme in Rhodesia now exceeds 100000 ha of highly productive wheat, maize, cotton, citrus and sugar production. There are 85 major dams, holding 3 km^3 of irrigation water, and over 7000 small dams already built in Rhodesia, with many more sites surveyed for development.

Not all irrigation schemes need massive capital and an enterprising example from the barren sandy coastal wastes near Asmara at the foot of the steep mountain escarpment illustrates this. Annually water pours down the escarpment and crosses the coastal strip in wide torrents. An Italian and Ethiopian partnership brought a small bulldozer in each dry season to spend three days in throwing up a small bund across the wide flat dry torrent bed. This deflected the first three or four floods into an adjacent flat flood plain area, where the water soaked in. Bigger floods then washed the bund away and salad crops were planted on the watered area. Small diesel pumps in shallow wells raised the stored water and irrigated the peppers, egg plants and tomatoes, which gave excellent crops.

Grazing lands

Since Professor Tribe will be discussing animal husbandry in the semi-arid tropics of Africa and Australia, I will deal mainly with effects on soil and water resources. Cattle-owning subsistence farmers keep large herds of low productivity, without regard to their efficiency in yielding meat, milk or money. Even African farmers entering the trading economy tend to keep excessive numbers on the communal grazing grounds. Irrigation schemes attract only the more enterprising from traditional tribal society, but they invest their profits in cattle on the surrounding areas. The result is soil erosion on a massive scale. Figure 1, plate 1 illustrates the state of thousands of square kilometres of southern Africa at the end of each dry season. Under good management the scrub is either slashed or cleared mechanically, and the grazing areas are closed in turn to provide tall dry grass which can be burned to destroy regenerating thorn scrub. Over-grazing leaves no grass to burn. Two experimental watersheds, each of 500 ha, in this condition in Uganda were instrumented and flumes were built to measure torrent flow. Water penetration was measured electrically on 20 profiles. Over four years, penetration of the annual 750 mm of rainfall reached a soil depth of only half a metre, and

FIGURE 1. Results of overgrazing in dry season. Thousands of square kilometres of savanna are reduced annually to this state in East, Central and Southern Africa.

FIGURE 2. Results of bush-clearing and two seasons without grazing. In spite of loss of more than 50 mm of topsoil these natural grasslands show great resilience. Fifty species of grasses were identified in this experimental catchment area.

(*Facing p.* 558)

40 % of all rainstorms was lost immediately as surface runoff. One valley was then cleared. Regeneration was rapid; (figure 2; plate 1) the rainfall penetrated 1.5 m into the soil and the stock-carrying capacity was approximately doubled. This and other examples are discussed in more detail elsewhere (Pereira 1973). Thus the soil erosion was not brought about by sheer numbers of stock but by lack of any form of grazing management.

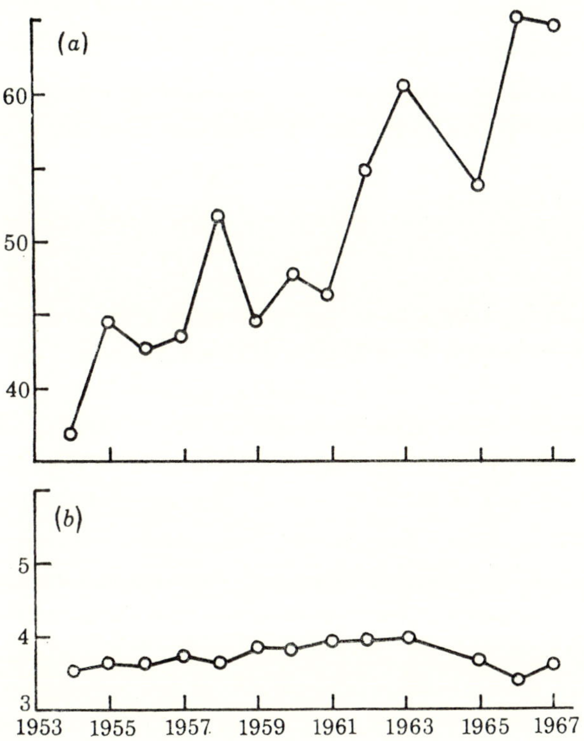

FIGURE 3. Progress in productivity of tribal cattle herds on communal grazing. (*a*) Total number of cattle sold. (*b*) Total number of cattle in Bantu areas (including Transkei but excluding SW Africa). (From Hamburger & Waugh 1968.)

However, the big annual variations in rainfall characteristic of the semi-arid areas cause major changes in the volume of feed produced. Carrying capacity can be reduced, by two successive bad years, to one-third of that of the previous five or more years. Recovery is rapid only if the grass is protected for the first few weeks of the rains. This faces the most sophisticated ranch management with very serious problems.

As early as 1923 the South African Government set up a Commission of Enquiry into Drought. A national Fodder Bank Committee reported in 1949. A major study and report on drought feeding followed in 1965. In 1967 an Enquiry into the Marginal Lands of Rhodesia studied the 1966 drought, in which 60000 cattle were moved by rail from the low veld to more favoured areas. Farmers are usually convinced that rainfall is dwindling, but the records, since the turn of the century, show no indication of decline; four or five successive seasons of rainfall well below the long-term average may occur by chance; such a drought can result in severe destruction of rangeland if stocking is not promptly reduced. The reports of the official Commissions show that fodder reserves are an essential provision for each ranching enterprise. Central fodder banks have been reported by successive Commissions to be expensive and ineffective. Maintenance of stocking below full level of herbage production in good years, and efficient off-take and marketing arrangements, are more essential in semi-arid country than in temperate climates. The problem can arise even in well-watered countries, such as Britain; we were seriously over-stocked in 1974/5. The development of offtake and marketing facilities in the Bantustans of South Africa is making encouraging progress, as illustrated by figure 3 (from Hamburger & Waugh 1968). Throughout southern Africa grazing control is complicated by the communal systems of land tenure, but their large scale offers great advantages for control, as is demonstrated on communal lands in western U.S.A.

A production constraint second only to drought throughout southern Africa is the incidence of both epidemic and endemic cattle diseases. Rinderpest, a major fatal epidemic virus, has been brought under control by large-scale vaccination campaigns, with the South African Veterinary Laboratories at Onderstepoort supplying vaccines to many countries to the north. Trypanosomiasis, eliminated from South Africa, is still spread by several species of tsetse flies over many thousands of square kilometres, and is held at bay in Rhodesia and in Kenya only by expensive 'fly barriers'. These are wide strips of country from which all bush is continuously cleared. The clearings, over hundreds of kilometres, are protected by expensive heavy wire fencing to exclude migrant elephant, buffalo, zebra, various antelopes, bush pigs, warthogs and other species; many carry the trypanosomes and most provide food for the tsetse fly. Steady progress in the containment of foot and mouth disease is encouraging, but quarantine measures have been necessary which sadly disrupt the offtake and marketing of stock. A serious tick-borne protozoan disease, 'East Coast fever', is controlled by weekly dipping of cattle. In sparsely roaded country this involves long weekly treks for thousands of beasts which leave wide scars of soil erosion for hundreds of kilometres. Soil erosion progressively spreads downhill and these trails cause major damage to semi-arid country.

The wildlife, which most countries in southern Africa are learning to respect for its tourist value, is a major complication for the veterinarian. In addition to the tsetse problem already mentioned the wildlife acts as a reservoir for other parasites and pathogens. Many studies and some commercial enterprises have explored the exploitation of the wildlife as a food resource (De Vos & Tecwyn Jones 1967; Talbot *et al.* 1965). Overall, these possibilities appear attractive but they have generated more discussion than profit. Eland have been successfully domesticated, but disease control is always likely to be a problem, since the veterinary profession has not yet the knowledge of the pathogens which affect wild species. A herd with which I was personally acquainted, however, had a better health record than most cattle enterprises (Posselt 1963). Eland certainly show very much better adaptation to marginal land conditions than do domestic cattle. The possibilities of game exploitation have been studied in southern Africa by a U.N. team (Swift, Pereira, Talbot & Beaton 1963) and some pilot schemes for harvesting wild species for fresh meat supplies to large industrial areas have provided encouraging practical experience. Costs have, as yet, remained too high for any widespread development (Pereira 1964).

International input of agricultural science

Successful experience by the Rockefeller Foundation in Mexico demonstrated that a major impact on the agricultural output of a developing country could be achieved by a vigorous team of international scientists, applying modern techniques of plant breeding, with management levels capable of reducing all of the main constraints on crop yields from the improved varieties. This was combined with training of Mexican agronomists, to use the new materials. As a result Mexico ceased major food imports and began wheat exports.

A joint effort by the Rockefeller and Ford Foundations launched a similar effort, with equal success, for the rice crops of SE Asia. Sponsored by the World Bank with two cooperating U.N. Agencies, F.A.O. and U.N.D.P., a consortium of national and institutional donors, the Consultative Group for International Agricultural Research (C.G.I.A.R.) now maintains a chain of eight International Centres around the tropical belt. C.I.M.M.Y.T. in Mexico (wheat and maize); C.I.A.T. in Columbia (cattle, pigs, cassava and beans for the vast barren Llanos of the Andes); C.I.P. in Peru (potatoes); I.I.T.A. in Nigeria (cowpeas, soya beans, yams, etc.); I.L.C.A. in Ethiopia (livestock systems of Africa); I.L.R.A.D. in Kenya (two major animal diseases of Africa); I.R.R.I. in the Phillippines (rice, both flooded and dry) and finally I.C.R.I.S.A.T. in India. The International Crop Research Institute for the Semi-Arid Tropics, set up three years ago at Hyderabad, is applying to sorghums, millets, pigeon peas, chick peas and groundnuts the process of assembly of a world collection of genetic material and breeding and selection under intensive pressure for suitability to the semi-arid tropics. A sophisticated programme aims at incorporation of resistances to many of the main diseases and even to some of the main pests, but the essential basis is rapid growth. Rain seasons which fail are characteristically of short duration, and early maturity is an essential stage in the reduction of the farmer's risks from an erratic climate.

It is necessary here to stress an issue which has been widely misunderstood and misquoted by economists and sociologists. While ability to respond to fertilizers is important, these new high yielding varieties (h.y.vs) have outyielded the traditional varieties even when no fertilizers are given, as a result of their ability to make better use of available moisture. Adoption of the new h.y.vs thus reduces the farmer's risks and does not involve him in extra costs unless he is enterprising enough to use fertilizers. They then enable him to profit from his enterprise.

The 30 donor-members of C.G.I.A.R. are helped by a small Technical Advisory Committee (T.A.C.) which works out priorities and policies for this very effective form of international aid. Since the crops of the semi-arid tropics have had far less attention than have the major world crops such as wheat and rice, substantial progress can be expected in spite of the harshness of the agricultural environment.

Conclusions

The low population densities shown by the maps of southern Africa do not imply that the continent could support vastly greater human populations. The overriding constraints are the scarcity of surface water and of potable groundwater supplies, together with the severity of the semi-arid climates for agricultural purposes. These sharply limit the areas suitable for intensive settlement. The fragility of landscapes whose geological rates of erosion are severely accelerated by man's activities of tillage, of livestock grazing and trampling, and even by his footpaths and roads, has already resulted in irreversible losses of soil resources over large areas.

The objectives of the Royal Society 'for improving natural knowledge' must still be applied, on the spot rather than from a distance, for nature is full of surprises and we have much to learn. But we are not 'starting from scratch' as all too many newcomers to Africa tend to assume. If the rapidly growing human population is to avoid many more disasters of the type experienced recently in the Sahel and Ethiopian famines, then there must be far more vigour in the application of what we already know and have fully tested. Contour farming, use of fertilizers and improved seeds, pest and disease control are the main issues for crop farming. Urgent reduction of livestock numbers by elimination of unproductive beasts and by rapid offtake of surplus cattle, is even more essential for the animal industry.

If the semi-arid country is to carry double the number of people in the next 25 years, the rural communities must give up their traditions of keeping large numbers of uneconomic livestock. Alternatively, if they are to preserve their traditions, then they must limit the increase of human populations. There is no way of carrying twice the population by the traditional destructive methods in this vulnerable country.

I do not believe that this essential change will be achieved, even with maximum international aid, until agriculture is recognized by each country's leaders as their most urgent concern. Until they allot rewards and status to food production which will attract the more able and enterprising, it will remain the occupation of the poor and ignorant, as it did for 1000 years of European history. Dr Boerma, speaking last year to the World Food Council as Director General of F.A.O., said that less than 10 % of international development assistance has been allocated to agriculture. This degree of attention will not secure the development of the resources of the semi-arid areas of Southern Africa to carry the human populations which they are now generating.

References (Pereira)

Anon 1965 *Report on drought feeding*. Dep. Agric. Tech. Serv., Govt. Printer, Pretoria.
Allen, W. 1965 *The African husbandman*. London: Oliver & Boyd.
Glover, J., Robinson, P. & Henderson, J. P. 1954 *Q. Jl R. met. Soc.* **80**, 602–609.
Griffiths, J. F. 1960 In *Tropical meteorology in Africa*. Nairobi: Munitalp Foundation.
Hamburger, H. & Waugh, W. F. 1968 *Proc. S. Afr. Soc. anim. Prod.* **7**, 29–40.
Lawton, R. M. 1963 *Kirkia* **3**, 46–82.
Manning, H. L. 1950 *J. Agric. Sci.* **40**, 169.
Pereira, H. C. 1959 *Nature, Lond.* **184**, 1768–1771.
Pereira, H. C. 1964 *Geogr. Mag.* **37**, 462–472.
Pereira, H. C. 1973 *Land use and water resources*. Cambridge University Press.
Pereira, H. C. & Thomas, D. B. 1961 *Emp. J. Exp. Agric.* **29**, 269–286.
Pereira, H. C. *et al.* 1961 *E. Afr. agric. For. J.* **27**. Special Issue.
Posselt, J. 1963 *Rhod. J. Agric. Res.* **1**, 81–87.
Rodier, J. 1963 *Bibliography of African hydrology*. U.N.E.S.C.O.
Swift, L. W., Pereira, H. C., Talbot, L. M. & Beaton, W. G. 1963 Report no. 63–44384. U.N. Special Fund. New York.
Talbot, L. M., Payne, W. J. A., Ledger, H. P., Verdcourt, L. D. & Talbot, M. H. 1965 *C.A.B. Tech. Comm.* no. 16.
Thomas, D. B. 1975 National College of Agric. Eng., Silsoe, U.K.
Trapnell, C. G. 1953 *Soils, vegetation and agriculture of N.E. Rhodesia*. Lusaka: Govt. Printer.
de Vos, A. & Jones, T. 1967 *E. Afr. agric. For. J.* **23**. Special Issue.
Walker, J. T. & Rijks, D. A. 1967 *Expl Agric.* **3**, 337.
Woodhead, T. & Waweru, E. S. 1970 *Expl Agric.* **6**, 51–55.

Discussion

H. C. Pereira, F.R.S. (Comment following mention of a field study in Botswana in which analyses had shown that browse from shrubs was higher in mineral elements than the grasses, and that this was related to the geomorphology of the subsoils). Minerals are important components of feedingstuffs but they do not keep starving cattle alive in the dry season. In Rhodesia, at Matopos, a large-scale comparison of cleared paddocks with heavily shrub-infested paddocks was conducted in a severe drought. When the sparse grass between the bushes was eaten, there was not enough energy supplied by the browse. The browsing cattle were reduced to acute starvation and had to be rescued, while those in the cleared paddocks needed no help. Some shrubs are indeed useful, but they can only supplement the grasses as a basic ration for domestic cattle.

On the question of the need for local study and measurement there is no dispute. Competent local scientific study for the detailed adaptation of techniques is a basic part of modern agriculture, as necessary in Britain as in Botswana.

T. Riney (*Department of Forestry and Natural Resources, Kings Buildings, University of Edinburgh, Edinburgh* 9). My question is based on Dr Pereira's emphasis on the importance of development and management of the headwaters of watersheds. In recent years in several areas of under 750 mm rainfall and in different continents, research and demonstrations have been made on raising the levels of water tables by changing the balance between woody vegetation and grassland in favour of grassland. Several successful examples are known from both the Western United States and Africa where it has been possible to control grazing, burning and other management practices. I would appreciate your opinion regarding the development potential this approach may have in under 620 mm rainfall areas with a recent history of lowering water tables and a present high proportion of shrubs or trees.

B. Dasgupta (*Institute of Development Studies, University of Sussex*). May I sound a note of warning to younger western scholars, that they might often find the poor countries which need technological help most, reluctant to receive such help. I would like to make two or three points to explain such reluctance on the part of the poor country governments. Firstly, it is often the experience of the l.d.cs (e.g. the implementation of the Bimas programme in Indonesia) that the technological help given from the West does not cater for the specific needs of a country given its environmental and social peculiarities. Secondly, often the foreign advisors recommend energy-intensive, capital-intensive, import-intensive technologies which are unsuited to the conditions of energy-scarce, low import capacity, poor countries with abundant labour supply. Thirdly, such technological help usually comes through the institutional form of multinational corporations which, because of their past colonial association and profit-maximizing motives, are distrusted by the governments of poor countries.

A. B. Rains (*Land Resources Division, Tolworth Tower, Surbiton*). Dr Pereira's inclusion of areas not usually considered to be in the semi-arid category is welcome. In the higher rainfall Sudan Zone of West Africa there is continuing environmental deterioration as the result of the activities of both the cultivator and the pastoralist; some of the latter's practices are particularly damaging. Throughout the zone there is also a largely uncontrolled exploitation of woody vegetation for fuel. Because of the large population in the Sudan Zone any decline in yields could be extremely serious.

Phil. Trans. R. Soc. Lond. B. **278**, 565–582 (1977) [565]
Printed in Great Britain

The conservation and improvement of resources: the grazing animal

BY D. E. TRIBE

School of Agriculture and Forestry, University of Melbourne, Parkville, Victoria, 3052, *Australia*
and
Program Committee, International Livestock Centre for Africa, Addis Ababa, Ethiopia

A comparison of livestock productivity in the arid zones of central Australia and sub-Saharan Africa is used to illustrate the ecological, social and economic factors which together determine the contrasting ways in which animal resources may be developed. Severe ecological restraints necessarily result in fluctuating animal populations and in the need for effective measures to control soil erosion, to prevent serious over-grazing, to vary stocking rates and to maintain the water supply. The aims, attitudes and needs of livestock owners also need to be recognized. Australian cattlemen primarily want to produce meat in order to make money; many African nomadic pastoralists primarily want to produce milk in order to feed their families. Degrees of achieving these aims and, at the same time, of preserving the necessary measure of ecological stability depend also upon a range of structural, economic and fiscal policies and opportunities. The need to coordinate livestock and rangeland research, planning and development on a multi-disciplinary basis is emphasized and the difficulties of doing so discussed.

INTRODUCTION

So much has already been written on this topic, so many experimental investigations have been reported, so many expert committees have made their recommendations, that one can only marvel that so little improvement has so far been achieved in conserving and improving animal resources in the drier parts of the world.

We can all recognize many of the pieces of the jig-saw puzzle which forms a picture of improved animal management in semi-arid zones. Some pieces, which are well-formed and illustrated in definite colours, are labelled: animal health, nutrition and breeding, water development, rangeland management, stratification of livestock industries, and marketing. Others, the shape and colour of which we can identify less clearly, are labelled: land tenure, agricultural credit, livestock taxation and fiscal policies, communications, community motivation, literacy and education. Finally, there are those pieces which we know are essential to the completion of the final picture but the outlines of which are still only vaguely seen and which are labelled: limitations to human and animal populations, social and political change, regional development based on ecological, social and economic criteria, and a reconciliation of national and international aspirations with the fixed and limited extent of the natural resources that are available. Despite our general familiarity with most of the various pieces, in no part of the world can we yet point to a finished picture in which all have been fitted into their correct place and which, as a complete work, supports within us feelings of long-term confidence and satisfaction.

The present paper attempts to contrast two necessarily incomplete pictures of animal management in semi-arid zones, one in Australia and one in sub-Saharan Africa. They are pictures of differing shape, texture and colour, yet both serve to illustrate the problems facing those who are trying to complete such pictures in such a way as to give lasting satisfaction to posterity.

It is dangerous to generalize about the characteristics of the semi-arid zones of sub-Saharan Africa and Australia and, particularly, I have no wish to over-emphasize their similarities. Clearly differences in, for example, latitude, altitude, geomorphology, pedology, vegetation and climate are substantial even within these zones, let alone between them. Any inter-continental comparison needs to be hedged around by so many limitations, qualifications and important exceptions that, in most contexts, such an exercise has little point. Moreover, because the movements of grazing animals do not recognize isohyets and the strict definition of ecological zones, it is inevitable that animal management should be considered in relation to what may loosely be called 'the pastoral zone' which, at one extreme, includes areas in which cropping may be practised with a reasonably high probability of success to the other extreme in which no cropping is possible.

Erratic rainfall and disastrous drought are ever-recurring themes over most of both of the vast regions of Africa and Australia under discussion. Annual mean rainfalls of between 200 and 800 mm, with variabilities larger than the world standard values, are characteristic, as are old and nutrient depleted soils, a shortage of surface water, an absence of large concentrations of people, and, with the exception of some mining operations and tourism, there is a dependence upon grazing as the principle industry. The major form of land use in the drier areas of Australia is the grazing of sheep and cattle and in Africa these species are commonly used together with goats, camels and donkeys. The ways in which these pastoral industries have developed and now function in Africa and Australia differ markedly, as do their levels of productivity, technologies and aims. Yet they share certain basic problems and both face an uncertain and difficult future.

Australia's pastoral industries

The development of grazing industries in Australia's arid and semi-arid areas is of comparatively recent origin. However, for more than 30000 years, those recently described by Blainey (1975) as 'the prosperous nomads' had adapted successfully to their dry Australian environment, even though they had not found it possible to change it. The logic of aboriginal nomadism has been described by Thomson (1939) who stressed that it was based upon a traditional understanding of, and intimate relationship with, the climate, the marine and insect life, the maturing of plants and the habits of wild birds and animals. Clearly the 'productivity' of the aboriginal nomads was inevitably low; they did not cultivate crops nor did they tend flocks or herds. Yet Blainey (1975) considers that there was customarily more food available than was needed, so that there was normally that vital margin of surplus which meant that drought was not necessarily associated with famine.

The key to this apparently stable ecosystem, in which man had apparently found his continuing niche, was the fact that aboriginal populations grew slowly if at all. Cultural curbs on population increase, such as infanticide, abortion, abandonment of the sick, aged and feeble, and warfare, were practised regularly by most, if not all, tribes. Occasional plagues or epidemics had an additional, but not an overriding, effect. In other words, the aim of the aborigines was survival and they had evolved a way of life which generally, and in the long term, enabled them to do this successfully by limiting their demands to the resources that were available.

The invasion and settlement of Australia by Europeans after 1788 inevitably led in much of the country to the replacement of the aboriginal cultures with the commercial cultures introduced by the new farmers and graziers. Survival and prosperity were now related to economic strength and their aim was understandably and unashamedly that of making money.

The key to their way of life involved, not population control, but a contrived increase in the supply of food and raw materials. The technological and scientific skills introduced to Australia by 19th century migrants, and later developed within Australia by her own 20th century scientists, resulted in levels of productivity and standards of living much higher than those achieved by Australian or any other nomads.

Much of the arid zone was aggressively occupied by the European settlers, who selected and fenced large tracts of land, often the best land, as individuals, partnerships and companies. Clashes inevitably occurred between these pioneer squatters, bewildered aborigines, and well-meaning but often uncomprehending colonial administrators. These were eventually resolved by a series of legislative measures which included the provision in some parts of Australia of recognized freehold occupation, while in other areas, notably large parts of the arid zones, medium to long-term leasehold arrangements were determined.

During the last 100 years settlement and property development has taken place during periods when market prospects have been good and seasons have been favourable (Barnard 1969). Inevitably, when adverse market and seasonal conditions have prevailed the pastoralists have been driven into debt, into exploitative grazing practices which have degraded range resources, and, in some cases, into selling up and seeking an alternative way of making a living.

Problems of Australian pastoralism

The hardships suffered by the contemporary pastoralists are largely caused by climatic and economic factors outside his or his government's control. They have been minimized by the infrastructure and services which have been built up by a succession of sympathetic governments, which have had the economic strength to translate sympathy into help. To this must be added the individual and concerted efforts of an educated and articulate grazing fraternity who themselves have taken appropriate managemental, economic and political steps to try to obtain whatever they have considered they have needed.

Thus, given the essential prerequisites of common commercial and social aims between livestock owner and government, and legalized forms of land tenure which have provided security for the pastoralist together with the essential basis for management initiatives, the grazing industries have now developed relatively high levels of productivity.

Livestock owners have introduced exotic species and strains of cattle, both *Bos indicus* and *Bos taurus*, and have developed their own improved breeds, such as the Australian Shorthorn and the Droughtmaster, by judicious crossing and persistent selection. They have developed water points, largely from artesian sources; used vast lengths of barbed wire fence to segregate flocks and herds according to age and sex; controlled pests and predators (particularly rabbits and dingoes); adopted appropriate methods to control animal disease and soil erosion; and have developed improved techniques of grazing management, particularly during drought.

At the same time they have been assisted by government action in, for example, developing communications. The rôle of the traditional cattle drover has now been supplemented by a network of airways, railways, and roads which facilitate the marketing of livestock and help to minimize the isolation of the people. The flying doctor service is well-developed as are the correspondence schemes for all levels of education. The radio, and more recently television, play a vital part in communications throughout the arid zones and most livestock owners have their own radio transmitter as well as receiver.

Underlying these and other similar developments are two general features of overriding importance.

Stratification and integration

First, there is the way in which the livestock industries of the drier country have been integrated with the Australian economy as a whole. The arid zones are essentially regarded as breeding areas which mainly produce store or immature stock for fattening in locations of higher rainfall. Similarly, during drought, stock are, where possible, agisted in more favoured areas or feed is transported back into the drought locations. Many livestock owners, particularly the grazing partnerships and companies, run properties in more than one climatic zone in order to maximize this flexibility.

In more general ways, also, the livestock owner in the arid zones is supported by the community as a whole. His contribution to the country's development and economy is generally acknowledged and he holds a proud position in much of the folklore and even mythology of Australia's young and largely urbanized society. Add to this his well-organized and successful political influence and one can appreciate the willingness of governments and the community to support the livestock owner with a variety of fiscal, taxation, drought relief and credit measures: measures which have been made possible by the successful economic growth of urbanized and industrial Australia and by her growing mineral and energy wealth and, in many cases, are rationalized by considerations of national security and the need to have an existing knowledge of, and chain of communications throughout, Australia's sparsely populated north and centre.

Research and education

The second important buttress against the uncontrollable vicissitudes of climate and markets (i.e. export markets) is the bank of ecological and economic knowledge and understanding that has been accumulated in Australia by various universities and state and Commonwealth instrumentalities, particularly the Commonwealth Scientific and Industrial Research Organization (C.S.I.R.O.) and the Bureau of Agricultural Economics (B.A.E.).

The extent of this knowledge is reflected in various scientific publications, notably the annual *Arid Zone Newsletters* and the periodic *Studies of the Australian Arid Zone* published by C.S.I.R.O., and various occasional publications and articles in the *Quarterly Review of Agricultural Economics* published by the B.A.E.

A great deal of research, often financed in part by levies raised from cattle and sheep owners, has now been carried out on such topics as the water metabolism, heat tolerance, nutrition, behaviour and reproduction of wild and domesticated herbivores, particularly ruminants, in arid zones (see e.g. Wilson 1974). The results of this work, complementing those of similar studies carried out in various laboratories and institutions in Africa, the U.S.A. and elsewhere, have done much to enlarge our understanding of basic biological principles and processes and have led to improved methods of animal management and breeding. They have also provided an increasingly firm basis of fact with which to resist those superficially attractive notions, which have been proposed from time to time for more than 100 years, that Australia would do well to abandon cattle and sheep husbandry in favour of kangaroo farming or the production of exotic species, such as the rabbit, llama, deer or some species of African game.

An understanding of physiological mechanisms important in tropical environments has been used successfully by C.S.I.R.O. scientists in evolving systematic breeding plans which have now led to the development of two new breeds of cattle, the Belmont Red (Turner 1975) and the Australian Milking Zebu (Hayman 1974), which have considerable potential both for tropical Australia and similar regions overseas.

The work of veterinary scientists has been remarkably successful in keeping out of Australia many of those diseases, such as rinderpest and foot and mouth disease, which play such a serious and limiting rôle in the cattle industries of many countries. More than this, veterinary research and policy have together eliminated some extremely important diseases, such as contagious bovine pleuro-pneumonia, which had been widespread in parts of the arid zone, and vigorous action has led to successful means of controlling various parasitic, tick-borne, reproductive and other diseases.

Perhaps of widest significance has been the research and survey work led by C.S.I.R.O. ecologists which has mapped and classified the natural resources of Australia's arid zones. Upon this information has been built a system of range inventory and suggestions for rational management which must constitute the basis for long-term conservation and use (see, for example, Perry 1976).

Together the developments and measures which have been mentioned have resulted in an apparently stable grazing industry which yields an annual turnoff rate (for cattle) of from 25 to 30 %, with annual mortality rates varying from 3 to 5 % in normal seasons, although they may be much higher in drought conditions (Haug & Hoy 1970; Reeves & Crellin 1972). The arid and semi-arid zones at present contain more than five million cattle and 30 million sheep and rates of return on capital invested are such as to suggest that these industries will persist in the future (Waring 1973).

However, the story is far from being one of unbridled success or reassuring certainty.

Future problems

The central question concerning the overall effect of development and management on the natural resources of soil, water and vegetation remains tantalizingly unanswered. It has been reported that arid zone vegetation has declined in productivity and stock carrying capacity (Newman & Condon 1969) and the perennial plant populations have been reduced greatly in many locations (Perry 1969). Moreover, it has been pointed out that economic pressures combined with the sheer riskiness of the country, make a pastoralist 'transgress the laws of safe husbandry in fragile semi-arid and arid environments even when he is reasonably aware of the dangers and anxious to implement the appropriate measures to avoid them to the best of his knowledge and ability' (Waring 1973).

The pressures on the arid livestock owner still include some of an institutional type. For example, where land is held under leasehold there is a reduced incentive for capital investment and, as leases have terminated, many properties have been subdivided before the leases have been renegotiated. Even large properties held freehold have been resumed by state governments and broken up for closer settlement by ex-servicemen and others who seem continually to yearn for a grazing property. Increasingly costs of production have gone up but returns have generally remained low and extremely variable. This has meant that adequate net profits cannot be maintained indefinitely on smaller holdings and pressures for restructuring the industry into fewer and larger ranches are mounting.

Survey results indicate that the financial position of pastoralists has generally declined during the last one or two decades and, for example, net indebtedness on sheep properties in the pastoral zone of Queensland has risen to an estimated $31 000 per property or $92 million in aggregate (Bain & Waring 1971). Some 19 % of these properties had debts exceeding 60 % of their market valuation and, according to another recent report (B.A.E. 1971 *a*), approximately one in eight of woolgrower debtors in pastoral Queensland are unlikely to be able to service their debts at foreseeable prices for wool, mutton and beef cattle.

Because of the marked fluctuations in climate it is inevitable that productivity, and therefore income, must also fluctuate widely. For example, the B.A.E. surveys of the pastoral sheep industry of Queensland have shown that net farm incomes commonly vary in successive years from + $50 000 to − $30 000. The effects of such variations need to be minimized by policies which allow the accumulation of sizeable cash reserves or off-farm assets to offset the unfavourable seasons, by appropriately long-term taxation and accounting procedures, and by effective policies for rapid destocking to minimize drought losses and restocking to maximize the potential of good seasons.

Because of the collapse of the beef export market and the low prices which have been paid for cattle during recent years, most cattle producers in the pastoral zone of Australia have been placed in a tight liquidity situation. The average level of capital employed per specialist beef producer in the pastoral zone amounts to over half a million dollars. This means that large amounts of debt have to be serviced and further capital still needs to be invested in herd, plant and fixed improvements if production is to be maintained (I.A.C. 1975).

It has been estimated by the Bureau of Agricultural Economics that of Australia's 35 000 specialist beef producers as many as 17 000 have 'doubtful viability prospects', and the Australian Government was recently advised by its Industries Assistance Commission (I.A.C.) that, as an immediate short-term measure, low interest loans totalling about $100 million should be made available to those cattle producers whose long-term prospects were thought to be reasonable (I.A.C. 1975).

In recent years increased attention has been focused on the plight of the aborigines and plans have been formulated for the development of aboriginal controlled cattle properties in northern Australia. However, here again the economic outlook is bleak and it was concluded that such enterprises could not be expected to show a sufficient return 'to repay over a 50-year period the costs of development, even if interest was not charged on the capital involved' (B.A.E. 1971 *b*).

An increasingly important problem is that of transporting livestock across the vast distances within Australia and of shipping carcasses across the world to Australia's export markets. Modern methods of transporting, processing and storing cattle and meat involve large inputs of labour and fossil fuels, both of which represent rapidly escalating costs. Unless these costs can be controlled the price of Australian meat on the world market will in time become prohibitive and production in the pastoral zone will necessarily decline.

Although much research has already been done, the present state of knowledge remains woefully inadequate in many vital areas, such as, weather forecasting, the assessment of rangeland status and trend, the regenerative abilities of plant communities under different management practices, the determination of ecologically and economically viable stocking rates for particular locations and the effects of mixed as opposed to single-species grazing.

Moreover, the central economic question of resource allocation in relation to national priorities is unlikely to be decided in favour of the pastoral zone. Even when financial returns in recent years have been generally adequate they have been far from generous and the evidence from B.A.E. surveys indicates that there is little opportunity for intensifying management practices or further utilizing knowledge that is already available concerning nutrition, reproduction and range management. The only modifications to current management policies which are likely to be introduced in the foreseeable future are those requiring little capital or current inputs.

If economic resources are to be allocated to the grazing industries there is greater scope for increased productivity, the use of improved technology, and economic gain in the high rainfall areas rather than in the arid zones. Although, during the period 1961 to 1972, the cattle population of the arid and semi-arid zones increased by approximately 25 % to slightly more than 5 million, in the non-arid zones it increased by over 100 % to 18.7 million. During the same period sheep numbers in the arid and semi-arid zones declined by about 5 million to 30.6 million while in the non-arid zones they increased by some 15.3 million to 132.3 million.

Such is the uncertainty about the best forms of future land use in Australian arid zones that suggestions are still being made that domestic stock should largely be removed and the rangelands used as a vast 'national park' for scientific and tourist purposes. However, the abandonment of such a large part of the country would create considerable social and economic problems and, in any case, the change from pastoral use to a national park would not remove the problems of long-term management. It has been suggested that if the economic viability of the grazing industries becomes doubtful, rather than abandon the range, 'society might be better off to share the cost burden of continued occupation with pastoralists if stable farming patterns could be established' (Waring 1973).

It has also been suggested that increased diversification of resource use is likely so that the vast areas now devoted to cattle and sheep grazing 'will become a spatial mixture of various types of land use including national parks, tourism, beef, carpet wool, sheep meat, and possibly karakul production, with a large part continuing to be devoted to the traditional wool industry' (Perry 1974).

It may be concluded that Australia has enjoyed many advantages in the development of her grazing industries in the pastoral zone.

The aims of pastoralists and governments have been commercially based. Their policies relating to management, marketing, land tenure, credit, taxation, drought relief, research and extension have all been designed, albeit with limited success, to strengthen the profitability of grazing properties and maintain it on a continuing basis. Further, with the advantages of high levels of education, capital investment and managerial skill within the pastoral industry, and the generally sympathetic support of a prosperous and expanding national community, it has so far been possible to maintain the industry on a saw-tooth path of prosperity which, on the whole, has managed to keep itself above the level needed for economic viability.

Yet, nevertheless, the jigsaw puzzle remains far from complete. Major parts of the picture remain to be filled in.

It is not yet certain whether or not the rangelands have been maintained above the level needed for long-term ecological viability. The best forms of land use for the future are still being debated and, although ecological considerations are recognized as vital, most observers regard those of an economic and social type to be even more complex and difficult to solve.

The key to the future conservation and use (i.e. management) of rangeland resources is the determination of appropriately conservative and fluctuating stocking rates. Economic pressures tend to encourage over-stocking, particularly in poor seasons and on smaller properties. No scheme has yet been proposed which ensures the appropriate adjustments of stock numbers and which, at the same time, is economically acceptable.

Africa's pastoral industries

Some of Africa's pastoralists are nomadic, while many more are transhumant or seasonally migratory. Unlike the aboriginal nomads, they are traditional herdsmen and shepherds with their own well-developed systems of management, but their basic aim is largely subsistence and, until the last 50 years, their ways of life had survived successfully, in apparent long-term equilibrium with their environment, for centuries (see, for example, Jacobs 1975).

For the most part the African pastoralist regards his cattle as milk rather than meat producers and as such they are the mainstay of his subsistence. However, cattle also represent capital which can be realized if necessary because of their value as meat. Because personal and family security depend on animals the pastoralist is keenly concerned with the maintenance of as high a population of stock as is possible (see, for example, Baker 1968) with sheep and goats being important as sources of meat and cash, and donkeys and camels for transport.

To some extent the African herdsman lives on the fringes of a monetary economy but livestock have a multiple value and can represent variable combinations of wealth, prestige and prerequisites of adulthood, marriage or parenthood, as well as being convertible into money value. In many places he is included in the market at least to the limit which he believes he can safely extract from his herd and, in some cases, to the limit of his immediate needs and desires. However, the vital value of livestock remains subsistence and, in large numbers, as an 'insurance' against the ravages of drought (Gallais 1967). Various observers have stressed the rationality of nomadic behaviour in the light of their basic aim of survival (Monod 1975; Haaland 1975).

A second key feature of the African pastoral tradition is that, although livestock are individually owned, grazing and water resources are not, so that their use tends to be opportunistic, favouring short rather than long-term objectives.

Problems of African pastoralism

The difficulties, and some would say impossibility of introducing more efficient systems of rangeland and animal management under traditional systems of land use have repeatedly been stressed (see, for example, F.A.O. 1967; Schaefer-Kehnert & Brown 1973). The problem has been well described by Jahnke & Ruthenberg (1974) in the following passage:

> The pastoralist is highly dependent on milk as a regular source of food for his family. Given the natural environment and the husbandry techniques the milk yield per cow can hardly be increased. As the family grows more cows have to be kept. The pastoral societies too experience a population explosion and there is therefore a constant need to increase cattle numbers. The impact of droughts and epizootics which used to regularly reduce the cattle population has been lessened by water development and veterinary measures so that herds are quickly built up to reach the carrying capacity of the land. The carrying capacity does

not, however, constitute a binding limit for the individual pastoralist since land is common property. As a rational man he will tend to maximize his share from the common land in favour of his private herd. As long as there is still a chance of a cow giving another calf and of a steer putting on another pound of weight it is worthwhile keeping the animals. This, however, means that as long as there is such a discrepancy between the responsibility (and interest) for the grazing and that for the cattle there is an inbuilt mechanism for overstocking and overgrazing with consequent destruction of the environment. This process is the quicker and the more dangerous the lower the natural potential. It assumes disastrous dimensions during abnormally dry years which serve to bring into the open the inherent mechanism of pastoralism for self-destruction.

The solution to this problem will in no case be found in training of the pastoralists unless one deems it possible to teach a man *NOT* to try to survive by maximizing his share from a diminishing resource, nor will the solution lie in any technical progress or innovation alone. Rather the situation calls for a change of the organization of production.

In a few African locations, European settlement during this century has resulted in a change in organization. Ranches have been established, often in areas with high potential, much as they were in Australia and with much the same history of antagonism and misunderstanding between the settlers, the indigenous peoples and the colonial administrators. These ranching areas, as for example in parts of Kenya, have followed much the same pattern as in Australia. Although they have not enjoyed the same benefits of an institutional and infrastructural nature, nor the same support and sympathy from governments and the general community, they have had some relative economic advantages in terms of land prices and labour costs. European ranchers in both Africa and Australia now face economic and ecological problems to which there are no clear or proven answers.

During the last decade, changes in the organization of some indigenous pastoral communities have also taken place, notably in Kenya and Uganda (see, for example, Pratt 1968; Jahnke, Ruthenberg & Thimm 1974; Nsubuga 1975).

Although it is too soon to be certain about their ecological or economic progress and viability, the early indications in Kenya are that, except for individual ranching schemes, these developments (i.e. group, cooperative, company and commercial ranches) are socially viable (Simpson 1973) and 'appear to be one of the economically more successful agricultural projects in Africa south of the Sahara' (Jahnke *et al.* 1974). The Ankole-Masaka ranching scheme in an area of higher potential in Uganda has been judged to be successful economically, socially and technically (Nsubuga 1975).

Techniques used for measuring economic returns from rangeland projects have been criticized because they frequently use parameters which do not fully comprehend the degrees of risk and unpredictability which are inevitably involved. It has recently been proposed that Monte Carlo simulation techniques might appropriately be used to estimate these risks and their consequential effects in rangeland projects (Gunawardena, Simpson & Wynne 1975).

The ecological consequences of ranching projects in Africa are still being studied and no definite conclusions are yet possible. It is certain that a conservatively stocked and well-managed ranch can conserve and even improve the range resources. For example, it was reported that Skylab photographs showed that the Toukounous ranch was the only island of vegetation in the whole of the drought-stricken country of Niger (Catinot 1974). However, there has never been any doubt that range and animal management is comparatively easy provided that there is no necessity, for economic or subsistence reasons, to maximize stocking rates all the time. Ranches maintained by governments or aid agencies, such as are found in many countries of Africa, have often been technically successful but, although their accounts are not usually made public, their economic viability appears doubtful.

Many of these ranches have served as breeding centres or as demonstrators of improved management. They have benefited from a substantial supply of information produced by the several African research institutions concerned with rangeland management and animal production and health. Much of the older research literature is located in libraries in Belgium, France, the Federal Republic of Germany and the United Kingdom, while F.A.O. publications contribute a valuable source of recent material, especially for the new African states. Among recent reviews and bibliographies which are particularly helpful are those by Barrés (1974), E.C.A. & F.A.O. (1972), E.C.A., F.A.O. & O.A.U. (1973), Reinhold (1971), Deramee (1971), F.A.O. (1972), I.E.M.V.T. (1973), Le Houérou (1973), Meyn (1970), Pagot (1971 *a*, *b*) and Pratt (1972).

However, for several reasons this accumulation of knowledge has had only a limited impact in practice. Extension services, particularly among nomadic pastoralists, are generally inadequate and, in any case, work in extremely difficult situations, frequently among conservative and illiterate peoples whose aims may be markedly different from those which underly the research that has been carried out.

Where there are communal systems of land use it is extremely difficult to introduce successfully such practices as improved grazing schemes or exotic breeds or crosses which may be readily susceptible to local diseases and infections. The cost of innovation is often beyond the capacity of pastoralists who may have little opportunity for, or understanding of, appropriate lending arrangements. Communal land is unacceptable as a security to credit banks which may also suffer from a lack of administrative control over the borrowers. National exchequers, also, may find it difficult to meet the initial and continuing costs which inevitably arise when changes in the rangelands gather momentum.

In the past, development in the arid and semi-arid zones of Africa has frequently been stimulated by the enthusiasms of particular individuals or organizations, sometimes technical or administrative, often political, who may encourage the introduction of, say, a new stock route, cattle market, field abattoir, grazing scheme, bore hole, dipping scheme, research station or bull breeding ranch. Such isolated and non-integrated proposals have usually failed to achieve their desired impact and sometimes they have even caused serious harm (Baker 1974), partly because of inappropriate technologies and administrative weaknesses, but mainly because of the failure to recognize that development is a continuous and complex process. Improvements may be technical (e.g. diseases control), social (e.g. reform of land tenure), or economic (e.g. credit facilities or improved marketing organization), but they must occur in an appropriate order in relation to one another if they are to have the desired result (F.A.O. 1967; Tribe *et al.* 1970).

Repeatedly, when recommendations have been made by teams of rangeland and animal specialists for the development of the livestock industries throughout sub-Saharan Africa, attention has been drawn to the need for integrated programmes in which steps to improve rangeland management, livestock movement and marketing, the control of stocking rates, and better extension activities should all go hand-in-hand with disease control and water development. However, the more complex developments, many of which involve social change and continuing high costs, do not have the political or administrative appeal of, say, schemes of water development or animal vaccination. Thus the full recommendations have often been postponed while isolated measures have been adopted because they have been regarded as easy and likely to lead to quick results.

Although such measures may help to gain the pastoralists' interest, and perhaps even their confidence, they do nothing to resolve the central problems of land tenure, credit and management. Indeed, the main consequence of improving animal health and making available more water in isolation has been to increase further the animal populations, the grazing pressures, the overstocking and the degradation of range resources. For example, in 1970 there were about 560000 users of the pastoral hinterland of the Delta in Mali compared with 200000 in 1968 (Gallais 1975).

Three factors which have together accelerated the present declining situation in the dry, pastoral zones of Africa are advances in medical services, advances in veterinary services, and water development. The result has been an increase in the human population beyond the level that the land can support (F.A.O. 1974*a*; Le Houérou 1975). In turn this has produced the need to maintain increased numbers of stock to provide subsistence for the greater number of people. The disastrous effects of the Sahelian drought from 1971 to 1974 were the consequence of the unprecedented build up of human and animal populations during the preceding decade. People and livestock used to die of hunger, water deprivation and disease, often soon after birth. The introduction of improved medical and veterinary services, e.g. vaccination programmes, enabled increasing numbers to survive into adulthood when the range resources, limited by serious, although not unprecedented, drought, could no longer provide subsistence for beast and therefore for man.

However, it is a misleading oversimplification to blame the veterinarian and the water engineer for the present problems of overstocking. Their work has afforded a new level of security to pastoralists by eliminating or reducing risks which previously decimated their flocks and regularly threatened their livelihood. This achievement has provided conditions in which pastoralists may now be prepared increasingly to change their traditional attitudes towards their stock, to restrict the size of herds and flocks, to recognize the importance of quality as well as quantity, and to trade stock surplus to their basic needs (Gallais 1967).

The essential problems posed in *Limits to growth* (Meadows, Meadows, Randers & Behrens 1972) are seen in their simplest and most brutal form in such regions as the arid and semi-arid zones and here, more clearly even than elsewhere, *Mankind is at the turning point* (Mesarovic & Pestel 1975).

The urgency of the problem is generally recognized and, during the last two or three years, there has been a spate of surveys, inquiries, committees and conferences to study the situation and recommend appropriate remedial action (e.g. De Wit 1972; F.A.O. 1974*a*, *b*; Boeckh *et al.* 1974; F.E.D. 1974; Rockefeller Foundation 1975; Ministére de la Coopération, Fonds de l'aide et de coopération, et Organisation pour la Mise en Valeur du Fleuve Sénégal 1975). These reports are valuable because together they provide a detailed and comprehensive coverage of the problems and of the steps that might appropriately be taken to alleviate them. Yet the completeness of their coverage and their detailed thoroughness themselves constitute something of a disadvantage.

Those who have had experience of such committees and conferences, particularly those of an international nature, will understand that their lists of conclusions and recommendations need to embrace and reflect a broad range of technical, socio-economic, institutional, national, political, disciplinary and, even, personal, considerations. Such comprehension tends to confuse because it is often difficult to identify agreed priorities and consequently 'the thrust and objectives (of livestock development projects) vary considerably and often are contradictory' (I.B.R.D. 1974).

A less complex, even an over-simplified, analysis can lead more directly to the essential core of the problem which is the urgent need to limit the future increase of human and animal populations in the pastoral zones and to encourage the movement of surplus populations to zones of higher potential.

Population control and redistribution

Of the first importance is the control and redistribution of human populations since if this is achieved the subsistence need for higher animal populations is then itself removed. Apart from stressing the need to develop and extend suitable means of birth control, and to urge that the necessary biological and social research and education should be initiated without delay, it is outside the scope of this paper to deal further with this topic.

The redistribution of pastoralist peoples in areas of higher potential has been taking place voluntarily in some locations for several years (Stenning 1974) and the problems of the pastoral areas cannot be solved without accelerating and extending this process (De Wilde 1967; Kraus 1965).

However, the populations in the higher rainfall areas are themselves increasing and, in consequence, areas of cultivation are being pushed further out into the pastoral zone.

Apart from encouraging policies of population control in all climatic zones, there are two other possible ways of accommodating additional numbers of agriculturalists and redistributed pastoralists. The first, for which there appears to be considerable scope, is to increase productivity, crop and animal, by means of improved technology in the high potential country. The second, is to find some means of making habitable, by both man and livestock, the vast areas of potentially good country now infested by the tsetse fly (*Glossina* sp.).

In this connection the work of the recently established International Laboratory for Research on Animal Disease (I.L.R.A.D.), with its headquarters in Kenya, will be of great importance. This laboratory has been established specifically to investigate immunological methods of controlling trypanosomiases and Theillerial infections. If vaccines can be produced for large-scale use in the field, or if the efficacy of drugs can be improved, a tremendous opportunity will have been opened up for resettling agriculturalists, pastoralists and their animals and so relieving the existing pressures on limited resources.

Exciting though this possibility is, it needs to be remembered that the development of trypanosomiasis controls is likely to be an extremely difficult and lengthy procedure and, even if accomplished successfully, the operations of resettlement in previously uninhabited country involve social, economic and management problems of great complexity. Moreover, in many parts of West Africa, even if trypanosomiasis is brought under control, there remains the important problem of onchocerciasis.

The multi-disciplinary approach

Although further research on many animal and rangeland problems is still needed, the accumulated results already reported from the several national and regional research centres operating in Africa now provide a substantial foundation of technical knowledge on which to base development projects. The depressing fact is that this knowledge is largely unused because it has not been contained in any overall context of production systems and further disciplinary studies of the traditional type are unlikely to be any more useful. For several years it has been increasingly obvious that there is a need for 'a whole new field of integrated research involving a study of the effect of development and of specific inputs on the social structure and habits of the people, on livestock performance and on vegetation and land use' (Pratt 1968). Phrases such as 'the multi-disciplinary approach to animal production systems', or something similiar, crop up again and again in the development literature of recent years (e.g. Tribe, Nestel, Pratt & Thomé 1973) but the meaning is not always clear.

The 'systems approach' can lead to a relatively precise analysis and definition of inputs and parameters, to simulation studies and the building of models which can be expressed mathematically (see, for example, Swartzman & Van Dyne 1972). However, in the context of African pastoralism mathematical precision may be illusory and the integrated study of particular ecological, economic and social systems, such as those of the Tanzanian Masai, the West African Fulani or the Afars of Ethiopia, is likely to be far more empirical yet still extremely useful. An interesting example of the use of a 'descriptive' systems approach has recently been published by Chudleigh (1976) for the small-scale dairy production systems in the high potential areas of Kenya.

The International Livestock Centre for Africa

The need for multi-disciplinary research concerned with the application of knowledge to the immediate and practical processes of development has led to the establishment by the Consultative Group on International Agricultural Research of the International Livestock Centre for Africa (I.L.C.A.), with its headquarters in Ethiopia.

The specific focus of I.L.C.A.'s programme is integrative research centred upon the multi-disciplinary study of existing animal production systems and the formulation of new or amended systems. These studies necessarily involve social anthropologists and economists as well as biologists and seek to:

quantify vital herd statistics, husbandry methods and animal performance in relation to the environment and to the material and cultural needs of particular societies;

establish relationships between the biological, environmental, social and economic components of production systems and identify points in the systems which are sensitive or resistant to change;

identify those specific topics in which a lack of knowledge limits development and in which disciplinary research is therefore a priority need.

In addition I.L.C.A. has given high priority to the establishment of a comprehensive and bilingual (French and English) documentation and information service. Finally, it is developing a capacity to organize seminars and training courses for those engaged in livestock research, extension, planning and production, in order particularly to increase regional competence in the multi-disciplinary 'systems' approach to livestock research and development (see, for example, I.L.C.A. 1976).

It is an ambitious programme and in itself constitutes an interesting experiment. No one should underestimate the difficulties of organizing and managing an international, multi-disciplinary team with pan-African responsibilities. However, the support of the donor members of the Consultative Group on International Agricultural Research and the cooperation and encouragement of African countries has enabled I.L.C.A. to start its work with enthusiasm and confidence.

Nevertheless, the ultimate goal of conserving and improving the animal resources of the semi-arid zones of Africa will not be achieved simply by research activities, however well-planned and executed these may be.

The problems facing African livestock planners are curiously similar to (though of a different order from) those facing planners in Australia – although for different reasons. In both places the ecological and economic limits to production are more closely set than has yet been generally acknowledged. Stability, in every sense, involves a balanced system of management in which the pressures of people and animals are kept in equilibrium with available resources. No matter which part of the world we are thinking about, the prerequisites for achieving such a system are first the knowledge and understanding that result from multi-disciplinary research and second the social and economic changes that result from enlightened community behaviour and courageous political action.

Many colleagues in Australia and Africa assisted in the preparation of this paper with helpful suggestions and constructive criticisms. Of particular assistance were Dr J. Pagot and his staff at the International Livestock Centre for Africa. However, the omissions, errors of judgement and other inadequacies belong entirely to the author.

References

Bain, A. A. & Waring, E. J. 1971 A survey of indebtedness of woolgrowing properties in pastoral Queensland. *Q. Rev. Agric. Econ.* **24**, 150–170.

Baker, R. 1968 Problems of the cattle trade in Karamoja: an environmental analysis. In *Ostafrikanische Studien* (ed. H. Berger). Nürnberg: Wirtschafts- und Sozialgeographisches Institut.

Baker, R. 1974 Famine: the cost of development. *The Ecologist* **4**, 170–175.

Barnard, A. 1969 Aspects of the economic history of the arid land pastoral industry. In *Arid lands of Australia* (eds R. O. Slatyer & R. A. Perry). Canberra: Australian National University Press.

Barrés, J-F. 1974 *Analytical bibliography on the Sahel* (French/English), FFHC Action for Development, FAO and World Council of Churches. Rome: FAO.

Blainey, G. 1975 *Triumph of the nomads: a history of ancient Australia.* Melbourne: Macmillan.

Boeckh, E., Bremaud, O., Dumas, R., Huhn, J. E. & Compere, R. 1974 *Etude sur la situation actuelle de l'élevage dans les pays du Sahel et des mesures de sauvegarde á envisager.* Bruxelles: Commission des Communautés Européennes (F.E.D.).

B.A.E. 1971*a* *Debt reconstruction and farm adjustment.* Canberra: B.A.E.

B.A.E. 1971*b* *Economics of land development: aboriginal pastoral properties in Arnhem Land and the Northern Territory.* Canberra: B.A.E.

Catinot, R. 1974 Contribution du forestier à la lutte contre la désertification en zones séches. *Techniques et Développement* **11**, 7–14.

Chudleigh, P. 1976 The use of classification and description of animal production systems in the formation of priorities for agricultural research in Kenya. *Agric. Systems* **2**, 1–19.

Deramee, O. 1971 *L'élevage des ruminants en Afrique au Sud du Sahara – bibliographie.* Bruxelles: Centre de Documentation Economique et Sociale Africaine.

De Wilde, J. C. 1967 *Agricultural development in tropical Africa* (2 vols). Baltimore: John Hopkins Press.

DeWitt, G. M. 1972 *Livestock development in Western Africa.* Abidjan: I.B.R.D. Permanent Mission in Western Africa.

E.C.A. & F.A.O. 1972 *Prospects for production, marketing and trade in livestock and livestock products in Eastern Africa to* 1985. Addis Ababa: E.C.A.

E.C.A., F.A.O. & O.A.U. 1973 *African livestock development study* (2 vols). Addis Ababa: E.C.A.

F.A.O. 1967 *East African livestock survey.* Vol. 1, *Development requirements.* Rome: F.A.O.

F.A.O. 1972 *Bibliography of marketing and market studies of livestock and meat in Africa.* Rome: F.A.O.

F.A.O. 1974*a* *Report of an expert consultation on the formulation of an international programme on the ecological management of arid and semi-arid rangelands in Africa and the Near East.* Rome: F.A.O.

F.A.O. 1974*b* *A survey of the problem of the Sahelian zone with a view to drawing up a long term strategy and a programme for protection, restoration and development.* Rome: F.A.O.

F.E.D. 1974 *Conférence élevage Sahel: compte rendu.* Bruxelles: Commission des Communautés Européennes (F.E.D.).

Gallais, J. 1967 *Le delta intérieur du Niger: étude de géographie régionale.* IFAN-DAKAR: Mémoires de l'Institut Fondamental d'Afrique Noire.

Gallais, J. 1975 Traditions pastorales et développement: problèmes actuels dans la région de Mopti (Mali). In *Pastoralism in tropical Africa* (ed. T. Monod). Oxford University Press.

Gunawardena, A., Simpson, Morag, C. & Wynne, A. G. 1975 *Risk analysis applied to rangeland livestock projections.* Discussion Paper Series, no. 21. School of Economic Studies, University of Leeds.

Haaland, G. 1975 Some economic and ecological aspects of traditional systems of livestock management. *Antropologiska Studier* **15**, 21–29.

Haug, N. F. & Hoy, R. J. 1970 The economic and financial position of graziers in arid south-west Queensland 1962–63 to 1967–68. *Q. Rev. Agric. Econ.* **23**, 222–243.

Hayman, R. H. 1974 The development of the Australian Milking Zebu. *Wld Anim. Rev.* (F.A.O.). **11**, 31–35.

I.A.C. 1975 *Report on short term assistance to beef cattle producers in Australia.* Canberra: Australian Government Publishing Service.

I.B.R.D. 1974 *Proposal for a cooperative effort to rationalize the livestock industries of the Sahel.* Washington: I.B.R.D.

I.L.C.A. 1976 *Proceedings of seminar on evaluation and mapping of tropical African rangeland.* Addis Ababa: I.L.C.A.

I.E.M.V.T. 1973 *L'embouche intensive des bovins en pays tropicaux.* Maisons-Alfort: I.E.M.V.T.

Jacobs, A. H. 1975 Masai pastoralism in historical perspective. In *Pastoralism in tropical Africa* (ed. T. Monod). Oxford University Press.

Jahnke, H. & Ruthenberg, H. 1974 Organisational aspects of livestock development in the dry areas of Africa. *Proc. 2nd Int. Seminar on Change in Agriculture.* University of Reading.

Jahnke, H., Ruthenberg, H. & Thimm, H. 1974 *Rangeland development in Kenya: a review of commercial, company, individual and group ranches.* Africa rural development study background paper. Washington: I.B.R.D.

Kraus, W. 1965 *Entwicklungspolitische Probleme der Weidewirtschaft im den Trockengebieten,* Stuttgart: Weide-Wirtschaft in Trockengebieten. Giessener Beitrage zur Entwicklungsforschung, Reihe I, Bd 1, 163–170.

Le Houérou, H. N. 1973 *Contribution à une bibliographie des phénomènes de désertisation, de l'écologie végétale, des pâturages et du nomadisme dans les régions arides de l'Afrique et de l'Asie du Sud-Ouest.* Rome: F.A.O.

Le Houérou, H. N. 1975 *Nature and causes of desertization.* Tucson: University of Arizona.

Meadows, D. H., Meadows, D. L., Randers, J. & Behrens, W. W. 1972 *The limits of growth.* New York: Universe Books.

Mesarovic, M. & Pestel, E. 1975 *Mankind at the turning point: the second report to the Club of Rome.* London: Hutchinson.

Meyn, K. 1970 *Beef production in East Africa.* München: IFO – Institut für Wirtschaftsforschung Afrika-Studien. Weltforum Verlag.

Ministère de la Coopération, Fonds de l'aide et de coopération, Organisation pour la Mise en Valeur du Fleuve Sénégal 1975 *Approvisionnement en viandes de l'Afrique de l'Ouest* (4 vols). Paris: SEDES.

Monod, T. (ed.) 1975 *Pastoralism in tropical Africa.* Oxford University Press.

Newman, J. C. & Condon, R. W. 1969 Land use and present condition. In *Arid lands of Australia* (eds R. O. Slatyer & R. A. Perry). Canberra: Australian National University Press.

Nsubuga, H. S. K. 1975 Appraisal of the Ankole-Masaka cattle ranching project in South Western Uganda. Ph.D. thesis. Kampala: Makerere University.

Pagot, J. R. 1971*a* *Natural pastures, forage crops and farming in the tropical regions of West Africa.* Maisons-Alfort: I.E.M.V.T.

Pagot, J. R. 1971*b* *Cattle diseases in French tropical Africa.* Maisons-Alfort: I.E.M.V.T.

Perry, R. A. 1969 Rangelands research and extension. In *Arid lands of Australia* (ed. R. O. Slatyer & R. A. Perry). Canberra: Australian National University Press.

Perry, R. A. 1974 Future of arid land vegetation under grazing. In *Studies of the Australian arid zone.* Vol. II. *Animal production* (ed. A. D. Wilson). Melbourne: C.S.I.R.O.

Perry, R. A. 1976 The evaluation and exploitation of semi-arid lands: Australian experience. *Phil. Trans. R. Soc. Lond.* B. **278**, 493–505.

Pratt, D. J. 1968 Range development in Kenya. *Ann. arid Zone* **7**, 177–207.

Pratt, D. J. 1972 *Selected bibliography on African livestock and rangelands.* London: Overseas Development Administration.

Reeves, G. W. & Crellin, I. R. 1972 Pastoral properties in the Alice Springs District (N.T.); their economic and financial position 1965–66 to 1969–70. *Q. Rev. Agric. Econ.* **25**, 27–49.

Reinhold, Bartha 1971 *Studien über Fragen der Zeburinderzucht in den Tropen.* München: IFO – Institut für Wirtschaftsforschung Afrika-Studien. Weltforum Verlag.

Rockefeller Foundation 1975 *International development strategies for the Sahel.* New York: Rockefeller Foundation.

Schaefer-Kehnert, W. & Brown, L. H. 1973 Economic and social aspects of animal production in relation to conservation and recreation. *Proc. 3rd Wld Conf. Anim. Prodn* 47–52, (ed. R. L. Reid). Sydney: University Press.

Simpson, Morag 1973 *Alternative strategies for rangeland development in Kenya.* University of Leeds: Rural Development Study No. 2.

Stenning, D. J. 1974 *Savannah nomads. A study of the pastoral Fulani of Western Bornu Province, Northern Nigeria.* Oxford University Press.

Swartzman, G. L. & Van Dyne, G. M. 1972 An ecologically based simulation-optimization approach to natural resource planning. *Ann. Rev. Ecol. System.* **3**, 347–398.
Thomson, D. F. 1939 The seasonal factor in human culture: illustrated from the life of a contemporary nomadic group. *Proc. Prehist. Soc.* 209–211.
Tribe, D. E., Curry-Lindahl, K., Pagot, J. R., Sokolov, W. & Smith, F. 1970 Animal ecology, animal husbandry and wildlife management. *Nat. Resources Res.* (U.N.E.S.C.O.), 123–145.
Tribe, D. E., Nestel, B., Pratt, D. J. & Thomé, M. 1973 *Animal production and research in tropical Africa: report to the Consultative Group on International Agricultural Research.* Washington: I.B.R.D.
Turner, H. G. 1975 Adaptation of beef cattle to tropical conditions: an Australian study. *Wld Anim. Rev.* (F.A.O.), **13**, 16–21.
Waring, E. J. 1973 Economic and financial constraints in the operation of livestock enterprises on arid shrublands. *Proc. 3rd USA–Australia workshop on rangelands reserach.* Tucson, Arizona.
Wilson, A. D. (ed.) 1974 *Studies of the Australian arid zone:* Vol. II. *Animal production.* Melbourne: C.S.I.R.O.

Discussion

W. E. Ormerod (*London School of Hygiene, WC1E 7HT*). Professor Tribe has suggested that a possible means of relieving pressure on the Sahel would be to settle the Fulani herdsmen in more humid zones to the south after large-scale tsetse eradication. Contact with tsetse and the losses through trypanosomiasis that result from this contact are, apart from starvation, the main restraint on increases in the Fulani herds. It is possible, if this restraint were removed, such being the demand for meat, that this zone would also become overgrazed without relieving pressure on the Sahel which is so attractive to the Fulani as a breeding area: consequent southward movement of the Sahara could result. This possibility has been considered in greater detail by Ormerod 1976, *Science, N.Y.* **191**, 815 (February 27).

S. Sandford (*Overseas Development Institute, 10–11 Percy Street, London W1*). The practice of always using or recommending as an insurance policy 'conservative' estimates of livestock carrying capacities in semi-arid areas is unsound. In areas of unreliable rainfall the only truly 'conservative' estimate is that no livestock should be carried at all. This, by *reductio ad absurdum*, showed the high cost of this sort of insurance policy in terms of production foregone.

Various speakers had claimed that adequate scientific and technical information already exists to improve productivity in semi-arid areas. But, in estimating livestock carrying capacity of the same area of rangelands in Southern Ethiopia, four of the most well-qualified people to make such estimates had differed by a factor not of 25 % but of 500 %. In the light of this sort of discrepancy, claims that adequate technical knowledge already exists cannot be justified.

E. B. Worthington (*I.B.P., c/o The Linnean Society, Burlington House, Piccadilly, London W1V 0LQ*). Throughout the symposium there has been very little mention of the indigenous grazing and browsing mammals which evolution has produced in both Australia and Africa, very beautifully adapted to the primary production of plants. How does their competition for food and immunity from local diseases relate to the cattle, sheep and goats, all of which have been introduced from Asia? Is there a place for the indigenous fauna in the food-producing industries of arid and semi-arid lands, comparable, for example, to the place of red deer in the Scottish highlands?

A. B. Rains (*Land Resources Division, Tolworth Tower, Surbiton*). Professor Tribe has mentioned the harmful results of developing water and providing veterinary prophylaxis in semi-arid areas when these have not been components of comprehensive schemes; he also welcomed tsetse eradication projects in areas of higher rainfall.

Because of a failure to follow up and regulate the subsequent use of many cleared areas, I believe that the results of successful tsetse eradication projects may be as disastrous as the introduction of these other measures has been.

J. Swift (*I.S.I.O., Sussex University, Falmer, Brighton*). Professor Tribe's description of increasing indebtedness among Australian cattle and sheep farmers invites comparison with rising indebtedness to village moneylenders and merchants among traditional nomad pastoralist communities in Iran, Afghanistan, Turkey and Greece. This has ecological consequences. People who are getting deeper into debt are likely to exploit the land more intensively, and eventually to overuse it. They are also more susceptible to famine.

This is a good example of the need for a proper understanding between economics and ecology. In particular I suggest we need to identify the economic and social conditions, like rising indebtedness, in which land degradation takes place. In this field information is seriously deficient.

I would suggest that there is evidence that some, not all, traditional societies set up ecologically sophisticated land-use systems. These have recently broken down because, as I suggested yesterday, the land-users are losing control over their economy and over their environment as a result of increasing dependency on outside economic and political processes; rising indebtedness is one example, and there are many others. If this is true generally, in order to reverse the degradation of the land, it is necessary to encourage new economic and social conditions which reduce dependency, and which increase effective power of economic and ecological decision-making at the level of land-users.

This suggests a new set of decentralized and low-level development strategies, very different from the capital-intensive, centralizing, high-technology development currently in vogue. Such new strategies might include simple technical measures like the water conservation and spreading techniques described by Dr Hadley this morning, but should cover the whole economic, social and ecological dimension of development.

D. E. Tribe (In reply to the discussion). The presence of tsetse frequently has the effect of conserving natural resources such as soils, water and vegetation. The lack of water development also has a similar effect. It is imperative that tsetse eradication and water development should always be seen as only one part of a comprehensive and rational plan for the long term use of natural resources.

There is no doubt that more information is required in the field of rangeland ecology and management. Nevertheless, it is essential that we should try and utilize as effectively as possible the considerable amount of technical information that is already available. It is sensible in our present state of incomplete knowledge to recommend the use of 'conservative' stocking rates and it is not logical to argue that 'the only true "conservative" estimate is that no livestock should be carried at all'.

It has been suggested many times that indigenous fauna, particularly the plains game of East and Central Africa, might be more efficient as producers of meat than introduced species such as cattle, sheep and goats. Unfortunately, every attempt to do this has so far failed. There are many technological problems concerned with wildlife harvesting, processing, transportation and marketing. In addition, the underlying ecological assumption that exotic species are less efficient than indigenous fauna deserves further examination. The present evidence is inadequate but some of it, at least, indicates that cattle, sheep and goats are in some ways better adapted to fragile environments (e.g. low nitrogen and high roughage intakes, drought, water shortages and climatic extremes) than are many indigenous species. The interaction between ecological and socio-economic factors which press upon pastoral communities deserves much more detailed investigation. I warmly support the views expressed by Dr Swift and assure him that the staff of I.L.C.A. is now studying the range of conditions to which he refers in the hopes of formulating effective new strategies of the type he has in mind.

Phil. Trans. R. Soc. Lond. B. **278**, 583–592 (1977) [583]
Printed in Great Britain

The conservation and utilization of water

By D. Rijks
Centre de Formation et d'Agrométéorologie et Hydrométéorologie appliquée dans le Sahel, c/o UNDP, BP 256, Niamey, Niger

There is a need for a quantitative assessment of the factors that determine the water balance of each part of an area, cultivated or under natural vegetation, and of their variability in time and space. These factors are rainfall, soil physical characteristics, potential evaporation and topography, the latter acting on surface and/or subsurface inflow and outflow.

Knowledge needs to be assembled as to how to handle the rainfall received in an area. This aspect covers infiltration rates, methods to temporarily impound water on the field, knowledge about rates of water use and water recovery in small-scale storage reservoirs.

The next step is to ensure a careful use of water through a judicious choice of the cropping and farming systems and by methods of land preparation and of the timing and choice of farming operations.

It is important to increase the efficiency of water use through the choice of crops and varieties, by elimination of non-water factors limiting optimal growth and development of crops, by adequate crop protection and by adopting suitable harvest procedures. Energy inputs into the farming system are to be adjusted to the desired level of production.

Among the socio-economic factors that affect the economic use of water, are matters of land ownership and availability of credit for low volume investment.

The definition of the correct level of water conservation during a given cropping season is dependent on the requirements of other components of the cropping system, notably of the production of crops on adjacent low lying lands, usually cultivated from the end of the rainy season onwards.

Out of season (November–February) irrigated crops on relatively small areas may contribute very significantly to agricultural production during a period when water requirements are relatively low.

Introduction

As in most other dry regions, the agricultural production systems in the semi-arid zone of Africa south of the Sahara, the Sahel zone, have always been characterized by an unstable equilibrium between the human and animal populations and the natural resources. In the Sudanian zone, where average rainfall varies from 400 to 800 mm per year, the equilibrium has fluctuated between abundant food supply and harsh conditions of life, but widespread starvation has not been a threat so long as only the 'resident' population was concerned. In the northern Sahelian zone, where average annual rainfall is less than 400 mm, the variation has been between reasonable food supply and near starvation, except perhaps for small groups of people living close to the three great river basins – the Niger, the Senegal river and the Lake Chad system. Conditions in this northern zone have given rise to widespread movement of people and animals in very dry years, which have often had difficult consequences in the 400–1000 mm rainfall areas to the south.

Although several series of drought years occurred earlier in this century (particularly around

1910–14 and in the early forties), only the drought sequence of 1968–74 aroused worldwide concern. This was partly perhaps because communication was better than before, and partly because population pressure was greater than before, so that after the humid period preceding 1968, marginal land, that would in earlier times have been left as a buffer for use in sequences of dry years, had already been taken into regular use. Thus the opportunities for movement within the northern Sahelian zone were more restricted.

The question of whether or not the climate is changing in the Sahel zone has been amply discussed. This is certainly of academic interest, but it is not a major immediate concern for the population living in the area. For those who depend for their living on the agricultural production of the semi-arid zone, the problem is to formulate a new technology of farming, adapted to the present demands of the population and the year-to-year and decade-to-decade fluctuations in meteorological and hydrological conditions that affect agricultural production.

A technology that aims at a more effective use of water, which is correctly recognized as one of the limiting resources, will necessarily have to take account of limitations in other resources, notably the availability of human labour, alternative forms of energy, and plant nutrients. A new technology is of necessity a package deal. It may be an instrument in the government's policies on integrated rural development, lessening the flow of people to cities that are insufficiently equipped to accept them. In areas away from the main river basins, it may have to be a technology requiring investment, which, though widespread, is small enough to be provided by farmers, groups of farmers or villages. Nearer to the main rivers technology requiring larger investments on relatively small areas may prove beneficial, if all other factors in the technological package can be made available in commensurate amounts.

Solar energy is abundantly available in the area. Other forms of energy are either not so readily available or have not been developed. Since vegetation is a major agent which converts solar energy into a utilizable form, it seems logical to promote the growing of plants that make effective use of the limited amounts of available water.

Water resources

Studies of the conservation and use of water resources should define the amounts of water that can be expected to be available, the demand for water, and the manner in which the available water should be managed so as to be most effectively used: in certain cases demand may be decreased or adapted to the supply, or the supply at any one point in time and space may be effectively increased.

The studies must embrace agricultural production in the widest sense, from occasional grazing in the 50–100 mm area to intensive irrigated agriculture producing several crops per year. Apart from grazing, the most widespread crop production systems are: raingrown crops, grown during the rainy season on the higher ground; flood plain crops, often planted towards the end of the rainy season on lower areas; and irrigated crops, using either loosely-controlled flood irrigation or more strictly controlled irrigation techniques using advanced technology.

As the first two methods are practised on many millions of hectares in the Sahel, even a small increase in yield per hectare from the present average grain yield of 300–500 kg/ha will significantly increase total agricultural output.

Hitherto, hydrological and meteorological information about these two systems of cropping has not been available until after the end of the growing season. It has occasionally been used for statistical analyses, which have sometimes indicated ways of improving the agricultural practices. Because the information is not transmitted or made available in real time, it cannot be used to guide government policy decisions or farmers' actions during the course of the season, for example in allocating resources for growing and harvesting crops and in decisions about crop protection measures.

The supply of water

All water in the semi-arid area south of the Sahara, whether it falls directly as rain or is derived from rivers, is brought there by the masses of humid southerly air which have their origin over the Atlantic ocean. The extent of northward movement of this southerly air is indicated by the position of the Intertropical Convergence Zone, which moves, with a time lapse of four to six weeks, behind the 'movement' of the Sun, between about 5° N in January and about 25° N in July–August.

Rain

For precipitation to occur, the vertical distribution of water vapour in the atmosphere must satisfy certain conditions. The underlying, relatively cool, moist layer must extend to sufficient altitude and the vertical temperature distribution above this layer must be such as to permit large scale vertical turbulence.

Early in the season, rain usually comes in isolated thunderstorms; later on, the rain-bringing mechanism is the 'squall line', extending over several hundreds of kilometres in a north–south direction and moving from east to west at an average speed of 40–80 km/h. The mechanisms by which these 'squall lines' acquire energy and water vapour are still far from completely understood.

In the semi-arid area south of the Sahara, rainfall is measured at about 60 synoptic stations, some agrometeorological and climatological stations and several hundreds of rainfall stations. The main preoccupation of the meteorological services is to maintain the quality rather than the quantity of observations. Records of daily and monthly totals are collected and published as tables and as maps by the Agence pour la sécurité de la navigation aerienne en Afrique et en Republique Malgache (A.S.E.C.N.A.) and the Office de Recherche Scientifique et Technique Outremer, Paris (O.R.S.T.O.M.). Frequency distributions of 10-day and monthly totals have been calculated and published by O.R.S.T.O.M. Few analyses of intensity and duration of rainfall have been published.

Rainfall totals and frequency distributions with time can provide a reasonable assessment of the rain régime in an area, but, if the landscape is not flat (especially in the drier areas), agriculturally very important differences may be associated with variations in microtopography. As a result of runoff, small slightly lower-lying areas may effectively receive water equivalent to double the average rain. Farmers in these areas make full use of these local conditions. Isohyets by themselves are not necessarily an exclusive means of assessing effective supplies of water.

Equally important is the duration of the rainy season. There is, of course, a certain relation between the length of the season and the monthly and seasonal totals, as demonstrated by a frequency distribution diagram. Moreover, it is often observed that smaller seasonal totals are associated with larger year-to-year variability. This is the main reason why the risks of agriculture are greater in the drier areas.

Traditional farming systems have tended to move southward and northward in sequences of drier and wetter years. On the one hand, this has led some population groups to move in accordance with this pattern, and on the other, it has led some sedentary population groups to master several technologies and apply them as required in different years. Such sedentary groups are found in the northerly zones of the great river basins, roughly between 15 and 17° N.

Local runoff

Local surface inflow and outflow is sometimes exploited in subsistence agriculture to grow crops outside the principal growing season. This technique allows farmers to produce additional crops, adapted to the seasonal changes in the temperature and atmospheric humidity régimes, and it spreads the work load of rural families and communities. It also diversifies their diets. It may be possible to gain locally important increases in agricultural production by a better selection of suitable soils for this type of agriculture.

Surface water

An important amount of water can be made available in the semi-arid zone south of the Sahara by using surface water more effectively. More attention should be given to surface water hydrology so that plans can be made to take advantage of these resources. Most existing systems of irrigated agriculture have been based on flood agriculture. Little secondary or tertiary irrigation has been practiced.

Surface water supply in minor tributaries is often abundant from the beginning of the main rainy season until December. The main streams generally have a plentiful supply of water in their lower valleys up to March. The times when water is available are evidently associated with distance downstream. At the end of the dry season all watercourses are almost completely dry.

Plans are being formulated to control and regulate flow on the Senegal and Niger Rivers, to supply water all the year round for irrigated agriculture. This leads to a definition of three future phases in the water economy of irrigated crops grown either in the cool period from November to February or in the hot period from March to June. In the first phase, before the rivers are regulated, water will be scarce during the latter season, for which crops which produce a large quantity of dry matter per unit of water evaporated will have to be selected. In the second phase, when river flow has been regulated but before all land suited to irrigated agriculture has been brought into cultivation, water will not be scarce in any of the three possible growing seasons in any one year. In the third phase, when the total area of land suitable for irrigated agriculture has been taken into use, a large ratio of dry matter production to water used again becomes imperative.

The water balance

In all studies of the conservation and use of water for agriculture, the essential factor is not the total amount of water used but the length of time during which water is effectively and adequately available for crop growth. The real need is therefore for knowledge of the water balance in a given area, on a given soil type and under a given crop. The calculation of a water balance not only shows the periods of shortages but also the periods of excess, knowledge which is essential in the study of the possibilities of conserving and using water. The correct calculation of water balances requires not only detailed meteorological observations, but also data on soil physical characteristics and knowledge of the physiology of the types of vegetation to be grown.

Ideally, water balance calculations should be made from actual observations of rainfall and evaporation over a series of years, rather than on means or calculated statistical distributions. The assumptions about crop and soil characteristics which are used in such calculations can be tested against observations of the quantities and distributions of water in the soil at different times, measured with a neutron probe, Bouyoucos blocks, tensiometers or gravimetric sampling.

Water balance data may lead to the formulation of agricultural practices that aim to store in the soil as much water as is useful (not necessarily the maximum that can be stored) and that will permit the excess to be discharged as controlled runoff, to be recovered and stored in lower lying areas for later use. This method, which is under study at the International Crops Research Institute for the Semi-Arid Tropics, Hyderabad, India (I.C.R.I.S.A.T.), among other places, can increase, and spread, the effective supply of water in semi-arid areas with a short wet season and a pronounced peak of rainfall.

Methods to increase water supply in this (and other) ways cannot be used efficiently without real-time knowledge of meteorological data. In many semi-arid areas of the world, the infrastructure to assemble this real-time knowledge has not yet been established. Such an infrastructure consists of a well-distributed network of weather observing posts (including weather radar where necessary) manned by qualified observers, equipped with a properly functioning telecommunications system, and feeding into a regional data analysis centre, that can transmit its conclusions rapidly and effectively to government services and farmers who need the information. The data analysis centre should also have rapid access to observations from weather and 'Landsat' satellites.

The data accumulated at such a centre can subsequently be analysed statistically, notably to determine the length of the growing season and the probability of early season or mid-season droughts. Such information will guide medium-term decisions on proposed modifications to, or inputs into, agricultural systems and practices, particularly when it becomes possible to forecast droughts.

The demand for water

In the driest areas, where the dominant agricultural exploitation is grazing, all water that falls as rain can normally be used.

In areas where arable agriculture is possible but the wet season is short, the demand for water is that of the main crops grown; millet, sorghum, maize, different varieties of *Vigna* (cowpea), vegetables and groundnuts. In areas where the wet season is slightly longer, farmers sometimes practise a form of intercropping, often with different sowing dates for the different crops, and the demands for water is more difficult to calculate.

The two most commonly used ways of estimating potential rates of water use for the calculation of crop water requirements are Penman's formula and observations from a class A pan. The empirical coefficients used in Penman's formula have to be determined at selected main agrometeorological stations in any one semi-arid area. 'Crop factors' for all major crops grown have been determined at many different locations and it seems that a working consensus has been reached.

While Penman estimates of crop water use can be applied to both rain-grown and irrigated crops, if the necessary precautions are taken, class A pan rates are more commonly used for irrigated crops only. A vast amount of well-tested work is available on 'crop factors' to be applied with evaporation rates from the class A pan. Both Penman and class A pan estimates of evaporation can be used to estimate evaporation from open water, lakes and reservoirs.

The use of Piche evaporation data has led to some very misleading conclusions, especially in regions where the vapour pressure deficit and temperature regime vary markedly with season.

The correction formulae which have been proposed to allow for this variation do not seem to have been widely accepted.

When interpreting the calculations of crop water requirements, attention should also be paid to the amount and the type of dry matter that is produced per unit of water used. Irrigated rice would seem to be inappropriate in semi-arid areas, except on land that would be flooded anyway. Wheat and sorghum appear to be better adapted. When crops are being selected, especially for cultivation under irrigation in the dry season, other agrometeorological factors, such as day-length, radiation and temperature régime must be taken into account, as well as, of course, non-meteorological factors affecting the farming system as a whole.

A special aspect of crop selection in semi-arid areas is the difference between the effects on determinate and indeterminate crops of an inadequate water supply. For the former, of which sorghum is an example, inadequate supply can lead to total failure, while for the latter it usually affects only the degree of success. In traditional agriculture, farmers have sometimes met this sort of difficulty by using genetically mixed material or multiline varieties. It should be possible to adjust water supply to indeterminate crops so as to obtain the largest possible ratio of economically useful dry matter per unit of water used.

Crop water demand is, to a considerable degree, also a function of other agricultural practices, such as date of sowing, spacing, weeding, intercropping and fertilizer régime. While intercropping is certainly being practised by farmers in semi-arid areas, little scientific work has been done on the assessment of the water balance under such crops.

The handling and effective use of water

A number of measures can be taken to make water more useful. In the case of rain, most of these are agricultural measures. In the case of surface runoff and flood water, these measures may also be in the field of rural engineering.

For rain, a first requirement is to maximize the rate of infiltration of the early rains, which fall when the soil profile contains little or no available water. Measures to facilitate infiltration vary from region to region and according to many factors, such as soil characteristics, stocking rates, movements of livestock, and previous crops. Such measures may include the preparation of the soil at the end of the previous season, stubble conservation, and careful control of burning. Early season runoff can be decreased by agricultural methods which individual farmers can undertake.

On the other hand, there are cases when controlled runoff during the peak of the rains is beneficial, both to avoid the leaching of plant nutrients and to collect water in lower areas for later use. On soils in which water infiltrates or percolates slowly, methods of land preparation that hold water temporarily on the surfaces of the fields may be beneficial in the early part of the season, before the soil profile has become saturated with water. Dates of planting, and spacing and weeding practices all affect the effectiveness of the water supply, especially early in the season.

To handle surface flow and flood water efficiently, the downstream area that is to receive the water must be fully prepared. Apart from basic agricultural measures, a system for forecasting flow and flooding, based on a river-flow model in which real-time hydrological observations are used, will permit human energy and agricultural resources to be allocated efficiently in handling the water – an operation which requires much labour in a short time and can easily fail if actions are not properly timed and coordinated.

There are several ways in which water can be more efficiently used than is the case at present.

(*a*) Choice of crops and varieties which will give large yields in a *life cycle of appropriate length.* Based on agrometeorological or hydrological data, crops and varieties can be chosen that can make full use of the length of the growing season that can be expected, at a particular location, say in 9 years out of 10 for food crops or 3 years out of 4 for cash crops. These possibilities should be examined jointly by agrometeorological services and national and international agricultural research institutions. Once the season's probable length has been determined and suitable varieties have been chosen, the cropping system, sowing dates, layout and preparation of land have to be designed or determined accordingly.

(*b*) Choice of crops and varieties with a large *ratio of dry matter production* (or *saleable product*) per unit of water used, and with a large growth rate at the stage of development when water is likely to be liberally available.

(*c*) Choice of crops or varieties whose *canopy architecture permits intercropping*, so that the combined water requirement of the two crops is adapted to the water supply that can be expected. It is important, in cases (*b*) and (*c*), to ensure *an optimum supply of plant nutrients*, especially *nitrogen, at the time when water is adequately available*. This may be done by intercropping or by applying fertilizer. The beneficial effect of intercropping may be increased if leguminous crops are chosen with root depths and crop architecture different from and adapted to those of the associated cereal crop, so that maximum use of incident radiation for photosynthesis can be assured.

(*d*) The choice of crops or varieties in which most, or even all, parts of the plants can be sold or used – *all-economic or multipurpose crops*. Examples of such crops are sorghum with a large sugar content in the stem, grain crops with a large ratio of grain to straw, or grain crops with straw that is palatable for cattle, such as wheat rather than rice.

(*e*) To increase the effective use of water, it is important to *protect crops* effectively. Seed treatment is a simple measure, readily accepted by farmers if properly organized. Protection of young seedlings against grasshoppers can also easily be organized (and moreover grasshopper outbreaks can be forecast if agrometeorological data are available in good time). Protection of crops in the latter part of the cycle is equally important. Agricultural measures and choice of genetic material may be important to lessen damage by birds and rodents as well as by insects and disease. In areas where meteorological conditions in the later parts of the season favour the outbreak of diseases or the increase of insect populations, the choice of shorter term varieties, even if their potential yields are rather smaller, may be advisable.

(*f*) *Harvesting procedures*, and the *infrastructure* for marketing and distribution, even over relatively short distances, should be improved in order to even out the differences in output due to variations in water supply from place to place and from year to year.

(*g*) When water is scarce, it can be more efficiently used if the *energy input into the agricultural system* is properly matched to the needs of the agricultural system. There are several periods when the demand for labour is large, notably when the land has to be prepared and rain-fed crops have to be sown and weeded, when the harvest of rain-grown crops coincides with the sowing of flood plain crops, and in the interval between the harvest of one irrigated crop and the sowing of the next one. This interval has to be as short as possible in semi-arid zones, where intensive irrigated agriculture may be based on a rotation of two or three crops per year (5 crops in two years).

Timeliness of sowing operations is often even more important for yield in tropical than in temperate zone agriculture. One reason is that changes in temperature are both rapid and large; another is that potential evaporation often changes in the same ways, which results in rapid changes in the amount of water available for establishing crops, particularly on soils which cannot store much water.

The exact method of matching energy requirements to needs will require detailed study in semi-arid regions. Animal traction may not be possible in those areas where the greatest demand for energy is at the end of the dry season, when animals are usually hungry and relatively weak. Individual farmers often lack the resources to invest in mechanical equipment. It may be possible to encourage the use of hired tractors, or of small farm implements like those which the International Rice Research Institute has promoted for small-scale rice farmers.

In some rain-fed areas, end of season land preparation, in readiness for the next season, may decrease the peak demand for energy and help the early rains of the next season to enter the profile.

(*h*) In areas where the topography is suitable, *impounded water* collected from the surface runoff *in months of excess rain* can be more intensively used. This requires cheaply constructed storage reservoirs, which could be built cooperatively by small groups of farmers or by villages. Studies intended to make this technique easily and reliably available to farmers are being conducted at I.C.R.I.S.A.T. It is probable that such techniques will be more easily accepted by farmers if the water collected can be used on their own land. Solar or wind energy might be used to return it at an appropriate time, to the land on which it was originally received. Applied at the end of the season, this water could permit crops or varieties with a longer growth cycle to mature, or allow the cultivation of a short-cyle catch crops, such as chick pea or pigeon pea.

The common emphasis on preventing runoff can be exaggerated. In some cases a certain amount of runoff may permit lower lying areas, perhaps in drier zones, to be taken into production. Sometimes alluvial soils lower down in a catchment may yield greater economic benefit per unit of water, than soils upstream. An extreme case is represented by the water-spreading systems of the Negev and parts of north and northeast Africa. In other cases the combined benefits from two crops, each grown with a slightly deficient water supply, may be greater than that of a single crop grown with a fully adequate supply of water. The difference in the phasing of the harvests of two such crops may also be important. An example might be an early upstream pasture crop combined with a later downstream small grain or grain legume crop, which would also provide palatable straw.

(*i*) Few exact studies have been made on the possible beneficial effects of *windbreaks* on the efficiency of water use of crops in the semi-arid zone south of the Sahara. Many side effects concerning, for example, the protection of crops against insects and birds, or the occurrence of excessively warm temperatures during periods when air movement is restricted, are associated with windbreaks. Detailed knowledge is needed before specific recommendations can be made.

(*j*) Although most nomads, whose animals graze range pasture, use very effective traditional *warning systems* to manage their pasture resources, real-time meteorological information and satellite observations may help to formulate information that could be disseminated in order to regulate pressure on natural grazing areas.

(*k*) Irrigated forage crops, grown during the dry season near the main rivers, and in the close neighbourhood of centres of population could help to maintain or even increase the live weight of animals raised on range pasture, in the period before they are sold, and so increase the effective return, to the producers, of the rain-fall range.

(*l*) Range can in some circumstances be improved by sowing leguminous species from the air, and resulting improvements in yield and quality could increase live mass gain. Where there are risks of overgrazing, aerial resowing may even include seed of unpalatable grasses, that would assure the continued presence of a vegetative cover, improve infiltration into the soil and reduce excessive erosion, for example around recently installed wells.

Conclusion

A number of agricultural techniques that can make the use of water more efficient are known and have been tested, but they are not yet widely practised. With appropriate investments and support services, a major extension effort might lead to rapid increases in agricultural output. Farmers generally judge innovations on their intrinsic profitability, but they will usually reject them if they increase their risks – for example (particularly in subsistence agriculture) by threatening the output of their food crops. Acceptable innovations may include relatively simple operations in seed treatment and crop protection. Information about such innovations should be more fully exchanged between anglophone and francophone countries in the drier parts of West Africa.

The collection, analysis, interpretation and dissemination of agrometeorological and hydrological information has received too little attention in many semi-arid areas in the past. Such information could facilitate both the detail and the timing of existing techniques of agricultural production in ways which could make the use of limited amounts of water more efficient. Weather forecasts, and particularly forecasts of the onset of rain and of mid-season droughts, are still far from perfect, mainly because we do not sufficiently understand many aspects of the dynamics of tropical climate and weather. However it may be that the analysis of weather at the 700 mbar level could help to improve forecasts.

On the other hand, we already possess the essential basic knowledge to forecast flows in streams and rivers, and are held back mainly because the necessary networks of meteorological and hydrological observations are neither complete nor efficient, and because we do not have fully satisfactory mathematical models for particular catchments and streams. Artificial rain-making remains a dream rather than a reality; the successes claimed so far seem to be induced accidents, rather than reproducible events. Some general weather lore exists in semi-arid zones, but little has been recorded, and still less tested or explained.

We know too little about the size and nature of the underground water resources in the semi-arid areas of Africa, to use them widely or intensively. Even under areas with little vegetation, water levels have dropped considerably during the droughts of the past decade. The present equilibrium of soil water is both unstable and precarious. Ill-considered and hasty use of ground-water resources on a substantial scale could have disastrous effects, which might be permanent by the time they became manifest. A comparison of water equilibria with grazing equilibria in the area might be instructive.

The rate of potential photosynthesis in the area is large. Given the required inputs of plant nutrients (notably nigrogen and phosphate), and of energy, it might be possible to make more efficient use of that part of the water resources that at present goes unused. This could lead to important increases in agricultural and notably in cereal production.

Discussion

R. A. S. Ratclifee (*Meteorological Office, Bracknell, Berks.*). The Meteorological Office has carried out some research aimed at forecasting whether summer rainfall in the Sahel zone will be above or below average. These methods use data from April and May and can give a forecast of summer rainfall category at the end of May. Would these be of value to Dr Rijks's organization?

D. Rijks. Yes.

R. E. Sowden (*The British Council,* 10 *Spring Gardens, London SW*1). We have heard today and yesterday that many of the technical problems of resource development can now be solved through the application of known scientific principles and techniques. We have also heard that effective implementation of known solutions is frequently difficult owing to associated socio-economic factors; in other words technological and financial inputs alone are not enough. I would therefore ask what further inputs in the broader field of education and training the industrialized countries can, and should, offer the third world in this respect.

Should the industrialized countries offer, for example, to provide more training of the MSc type for deskbound administrators so that they can better understand the full implications of their plans, or 'on the job' training for professional practitioners or even, dare I say, practical training for the planners? Alternatively one might, perhaps, concentrate instead on the provision of training for instructors and so utilize the multiplier effect, or devote more of the available but limited resources to the training of the local practitioner, the technician on the job.

D. Rijks. There are two bottlenecks in the data collection interpretation and dissemination programme that necessarily need to be solved on a national level. The first is the training and the employment of observers and technicians that assure the flow of basic data towards the data interpretation centres. The second is the training and the employment of technicians and field staff that ensure that the information, formulated by national hydrological and meteorological services, arrives in an understandable form at the farmer's level and can be implemented by him. There is thus a need for training of this field staff and the technicians, and for governments to make the necessary funds available for them to execute their work.

C. J. Bevan (*U.L.G. Consultants* (*Warwick*) *Limited,* 2 *Church Street, Warwick, Warwickshire*). We have heard that Dr Rijks and his colleagues intend determining crop water requirements using the Penman method. While using this method in an arid zone may lead to some controversy, it will undoubtedly be of great value to agronomists and others to have information of this kind available for the crops of the Sahel.

I would like to know what system of nomenclature, and which symbols, Dr Rijks will use for evaporation and evapotranspiration measurements. Travelling throughout the world a multiplicity of expressions and symbols are found, and much time is wasted in converting such expressions back to more familiar notation. It is high time, surely, that an international terminology for this subject was agreed upon.

D. Rijks: Our group will, understandably, use the terminology recommended by the World Metereological Office.

Phil. Trans. R. Soc. Lond. B. **278**, 593–609 (1977) [593]
Printed in Great Britain

A weather watch for semi-arid lands within the tropics

By J. Elston and M. D. Dennett
Department of Agricultural Botany, The University of Reading

Meteorological measurements, at the Earth's surface and in the atmosphere, and satellite photographs are exchanged internationally. They can be used to follow the daily and monthly behaviour of the climatic variables that affect agriculture. The consequences of variation can be assessed to some extent by using heuristic models of crop growth and to a lesser extent from statistical correlations with yields.

The movement of air masses within the tropics has a systematic but irregular pattern, which is still not well understood. However, the statistical correlation of some weather variables, such as geopotentials, with the rainfall to come, can be used to indicate the character of the coming season in some parts of the tropics.

1. Introduction

Agricultural systems are determined, to a large degree, by climate and weather. Meteorological measurements are made throughout the world, and three main sorts of meteorological informal tion are exchanged internationally. Measurements made several times daily, at the surface and in the atmosphere, are sent on a complex telecommunications network almost continuously to form the 'World Weather Watch' organized by the World Meteorological Organization. Copies of satellite photographs, taken both in the visible and infrared wavebands, from the National Oceanic and Atmospheric Administration of the United States of America are sent on a facsimile transmission system, usually daily. The monthly means of the measurements are sent out, early in the following month, on the 'Climat' system of the World Meteorological Organization.

This international exchange would be remarkably complete if it all worked all of the time. Even in the arid regions of the tropics there is a reasonable spread of meteorological stations which could report measurements at the surface (see figures 1 (west and northwest Africa) and 2 (Indian subcontinent)), though there are inevitably many fewer stations for atmospheric measurements. Unfortunately a number of countries almost continually fail to report, and others occasionally fail to report. More rarely, there are telecommunication failures.

The kinds of measurements which are transmitted are primarily determined by the needs of synoptic and dynamical meteorologists. Fortunately these include most of the measurements at the surface and in the atmosphere that are important in agricultural meteorology. Probably, means over periods of 5–10 days are the most useful starting data for agricultural meteorology because this is the length of period within which a crop changes measurably. However, it is far from easy to collect and derive 5-day means from measurements which are made daily, or even more frequently, for a very large number of places. We have therefore had to start, for lack of anything better, with the monthly means from the 'Climat' network. The use of periods as long as this may obscure important statistical and physiological relationships between weather and yield.

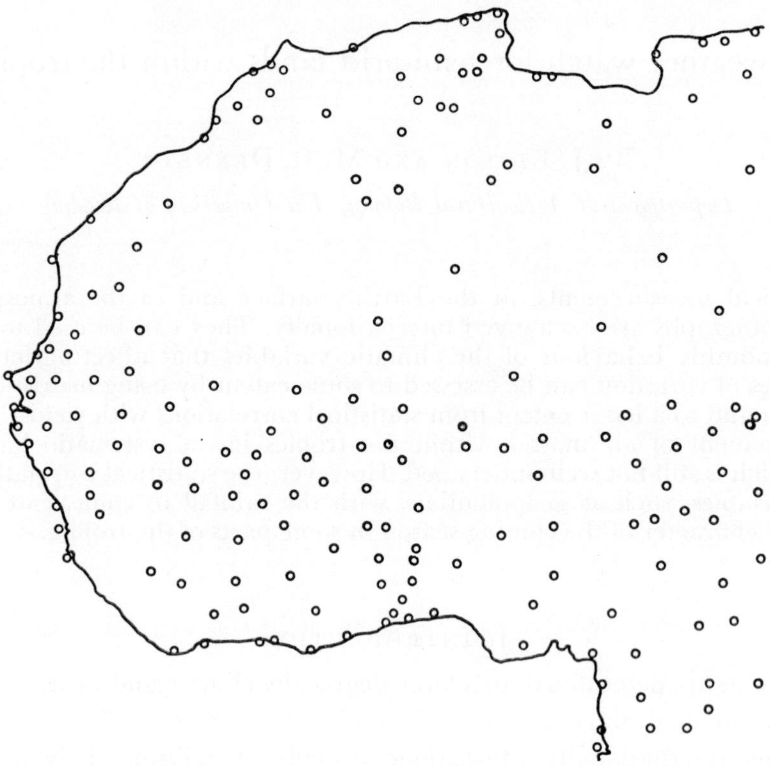

FIGURE 1. A map of West Africa showing the meteorological stations which could report surface measurements to the 'Climat' network.

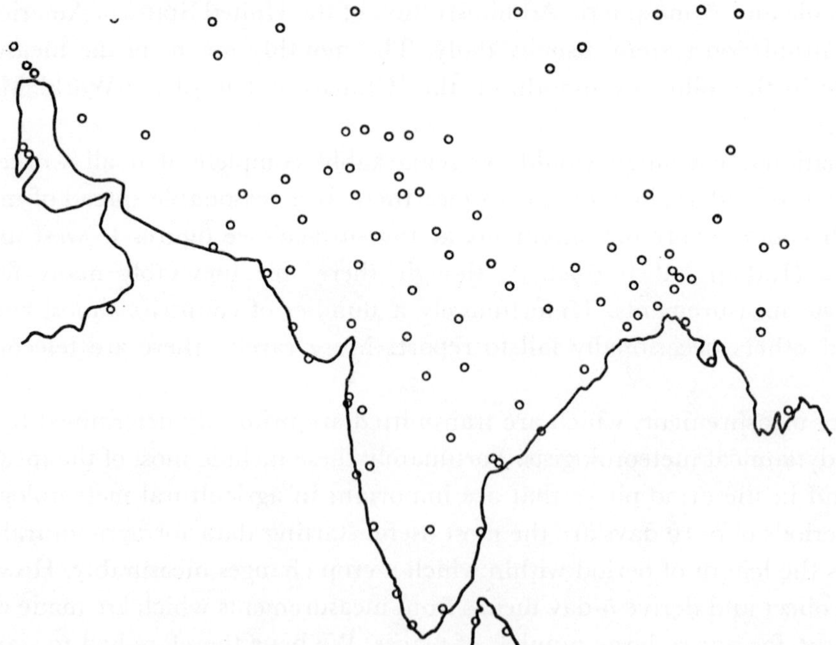

FIGURE 2. A map of the Indian subcontinent showing the meteorological stations which could report surface measurements to the 'Climat' network.

2. Models

The meteorological measurements must be organized within some conceptual frame before they can be used for most agricultural purposes. The development of such a frame is now one of the principal objectives of agricultural meteorologists throughout the world. Both statistical and heuristic models have been made to relate the yields of crops to weather variables. Such models are intrinsically more difficult to make for animals, partly because their behaviour is more difficult to quantify and analyse, but also because it enables animals to respond more flexibly to variations in the weather. Consequently, even though heuristic models relating animal production to weather would be very important for the arid tropics, no fully satisfactory ones yet exist (but see Gates & Schmerl 1975).

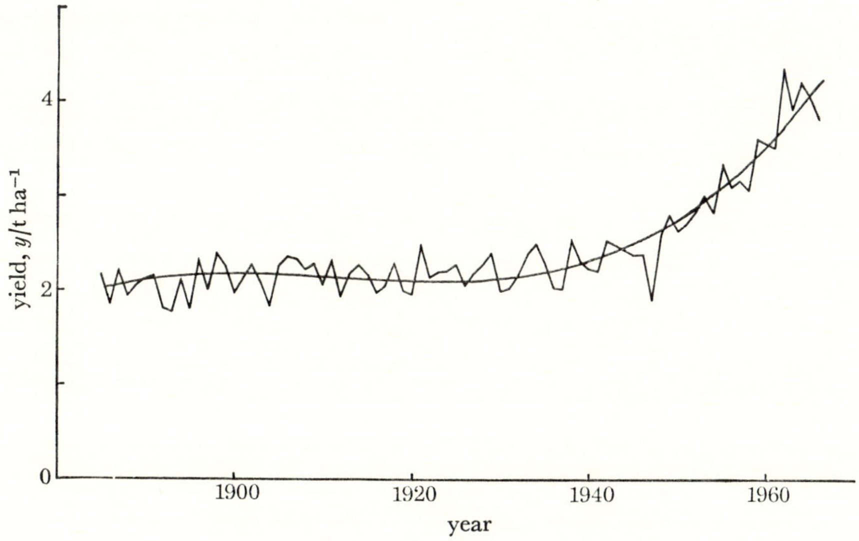

FIGURE 3. The yields of wheat in England and Wales from 1885 to 1966, with a fitted cubic regression ($r^2 = 0.89$).

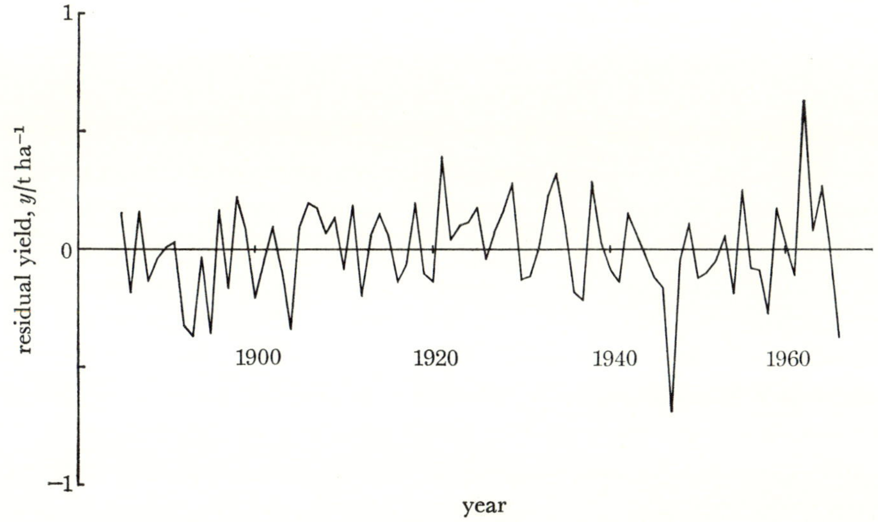

FIGURE 4. The residual variation of yield, after removing the trend, of wheat in England and Wales from 1885 to 1966.

A number of statistical models have been made to relate the yields of crops, often of cereals, to weather. Yields vary from year to year. They generally tend to increase with time; perhaps because agricultural technology improves and spreads. Polynomial regressions of yield on time can account for much of the variance in a long run, which might be more than 50 or more years of measurements. The fitted line can be taken to represent the effects of changes in farming practice and improved technology, though it is bound to include also the effects of any long-term variation in climate.

TABLE 1. THE MEAN YIELD, ITS STANDARD DEVIATION, AND THE STANDARD DEVIATION OF THE RESIDUALS OF YIELD, OF WHEAT IN ENGLAND AND WALES FOR 10-YEAR PERIODS FROM 1886

period	mean yield y /t ha^{-1}	standard deviation S_y/t ha^{-1}	coefficient of variation (%)	standard deviation of residual S_r/t ha^{-1}	standard deviation of residual as a % of mean yield
1886–1895	1.99	0.17	8.31	0.18	9.03
1896–1905	2.15	0.18	8.35	0.18	8.38
1906–1915	2.21	0.13	6.05	0.13	5.80
1916–1925	2.16	0.17	7.85	0.17	7.97
1926–1935	2.23	0.17	7.71	0.16	7.11
1936–1945	2.32	0.19	8.31	0.16	6.84
1946–1955	2.71	0.38	14.09	0.25	9.17
1956–1965	3.66	0.47	12.80	0.25	6.86

The average yields of wheat in England and Wales from 1885 to 1966 are given in figure 3, together with a fitted cubic regression. In this example yields have increased from about 2 to nearly 4 tonnes ha^{-1} during 80 years. The residuals from the regression line are set out in figure 4. They may be taken, at a first approximation, to represent the effects of weather or of the immediate consequences of variation in weather, such as fluctuation in the damage done by pests or pathogens. Statistical tests can then be used to correlate these residuals with variation in weather, especially solar irradiance and the water balance. The variance of the residuals itself varies with time; the standard deviation ranges from 0.13 to 0.25 t ha^{-1} (table 1); so that linear regressions are not always appropriate.

These models are statistical and not general. The equations relating weather variables to yield change from place to place. Even when the types of equation are similar, the constants vary. Furthermore, there are few long sets of reliable measurements of both weather factors and yield from which equations can be established, particularly in the tropics, though there are a few for industrial crops.

For example, records of sugar cane yield from Viti Levu, Fiji, have been collated by Dr P. C. Prasad (figure 5). The crop is grown on the seasonally arid western side of Viti Levu; not the most favourable environment for sugar cane. The yields have varied from 41.2 to 85.1 t ha^{-1}. The slope of the linear regression is small (0.3 % per year) and the yields are not generally correlated with either solar irradiance or water deficit (Prasad, Dennett & Elston 1975). The residuals are proportionately larger than for wheat, reaching 20 t ha^{-1} (figure 6).

The Food and Agriculture Organization of the United Nations has published measurements and estimates of yield for much of the world since 1948 in the Production Yearbooks. This is about the shortest period worth examining. However, the figures from the semi-arid tropics are sometimes inconsistent or unreliable.

A single weather factor of paramount importance may sometimes be statistically related to yield. Lomas (1972) found an asymptotic relation between rainfall and wheat yield in Iran. Kibukamusoke (1958) has shown that the yields of seed cotton in Uganda can be estimated from rainfall; but the form of the relation differs widely between different parts of Uganda.

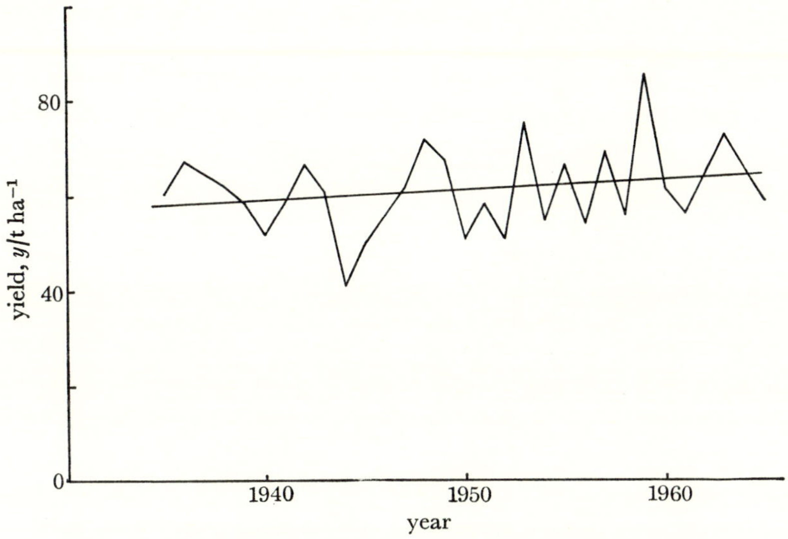

FIGURE 5. The yield of sugar cane in Fiji from 1935 to 1965, with a fitted linear regression (r^2 = 0.04).

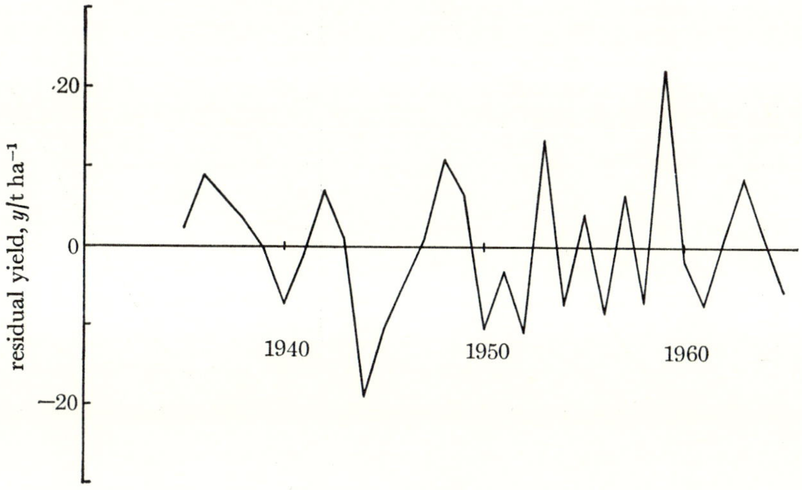

FIGURE 6. The residual variation of yield, after removing the trend, of sugar cane in Fiji from 1935 to 1965.

Heuristic models of growth and development, leading to general models of yield, are now being developed for crops (see Elston & Monteith 1975). The quantitative attributes of the crop at a particular time, such as the leaf area, the dry mass or the economic yield, depend upon the sum of the rates of many physiological processes. These processes have mechanisms, usually biochemical, whose rates are altered by meteorological variables. If the parameters that describe this dependence are known and if there are measurements of all the relevant meteorological factors then the states of the attributes can be predicted.

The rate of production of dry matter depends upon the solar irradiance intercepted by the crop. Monteith (1965, 1972) has shown how the physical structure of the canopy can be combined with the light response curve to predict the crop growth rate from the solar irradiance. This model, and variations on the same theme by other modellers, are accurate: figure 7 compares estimated and measured growth rates in four peanut crops at Samaru, Nigeria. The

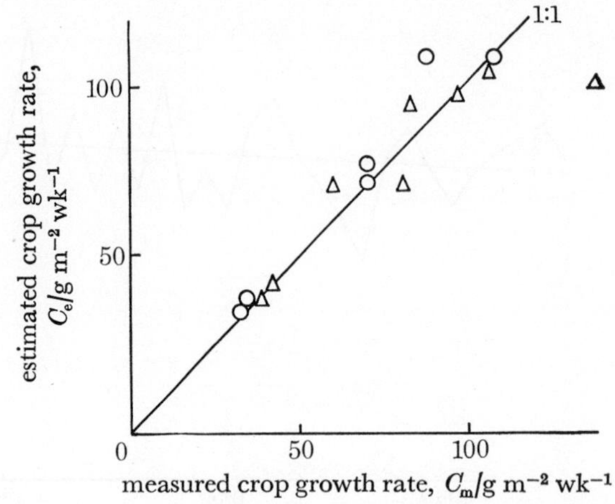

FIGURE 7. A comparison of estimated and measured growth rates for four groundnut crops from Samaru, Northern Nigeria. Taken from Monteith (1972).

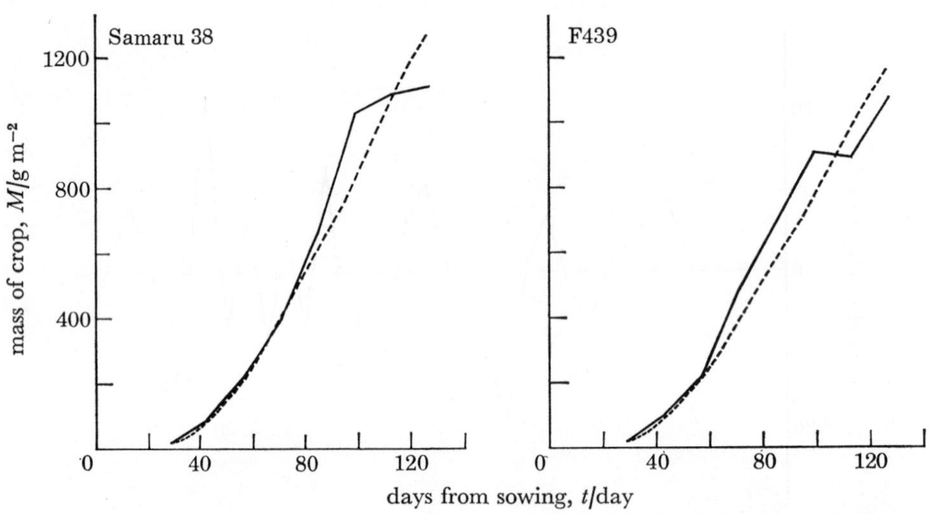

FIGURE 8. Prediction of the dry mass of an alternate (cv Samuru 38) and a sequential (cv F439) peanut crop based upon Monteith (1972) (dashed line) and the actual dry mass (continuous line) taken from Elston, Harkness & McDonald (1976).

points fall close to the 1:1 line. We understand that part of the *place/weather/crop* system that describes photosynthesis reasonably well, and that relationship is general. The estimates of the crop growth rate can be integrated over time to give an estimate of the dry mass of the crop. In the two peanut crops which provided the data of figure 8 the estimate of the final dry weight exceeded the actual mass by less than 20 %.

Unfortunately we do not understand sufficiently well the relations between environmental factors and respiration, leaf growth and death, or the partition of dry matter. As a consequence a series of empirical relations have to be used to model these processes. For the peanut in a number of seasonally arid climates the partition of new dry matter between leaves and the rest of the plant (the leaf to total growth ratio) is a linear function of time (figure 9), falling from 0.6 at about 2 weeks to less than 0.2 at about 15 weeks from sowing. Increases in leaf area can then be simulated. A similar partition coefficient, the reproductive to total growth ratio, can be used to model yield. However, disease and age affect or kill peanut crops in very uneven ways, so the predictions of economic yield are generally less impressive than those of total dry matter.

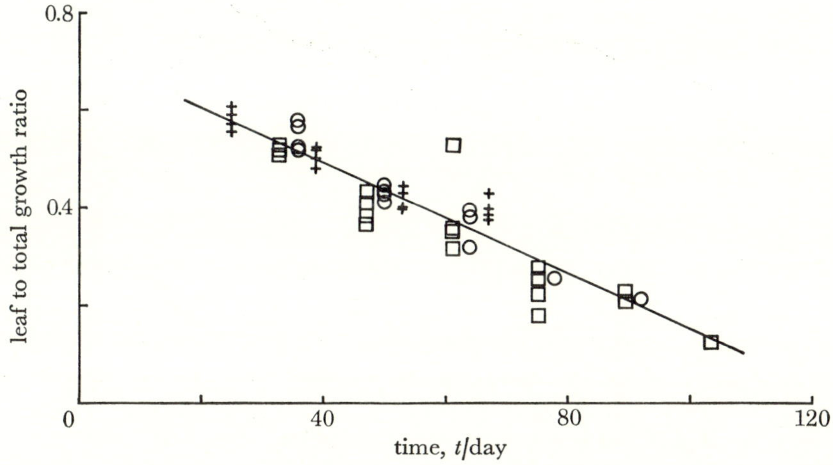

FIGURE 9. The leaf to total growth ratio against time for 13 crops of peanuts in three experiments at two places including six different cultivars with two different branching habits.

Similar empirical partition coefficients, though without the same simple dependence on time, can be made for other annual crops. The heuristic models do not at present describe development (as opposed to growth) satisfactorily. Indeed they remind us of all that is not known about weather and crop growth.

Many statistical and heuristic models, with weather variables as inputs, have been made. They are continually being used and some are being tested. They are still imprecise, but they do describe the limits of biological yield as a function of weather. The agricultural consequences of current weather will be more directly and reliably assessed as these models are improved.

3. The assessment of the current weather

A rising mass of warm damp air forms the meteorological equator. Surface winds from both hemispheres converge there to form the Inter-Tropical Convergence Zone. This moves in a regular fashion with time, bringing rain to the seasonally arid tropics. The basic physical system has long been understood (Hadley 1735). Both the mean position of the zone in any one month, and the mean pattern of rainfall, are known, but the movement varies from year to year. Moreover, a large part of the rainfall is convective, and is related to mesoscale disturbances within the convergence zone. Consequently the rainfall at one place can vary substantially from year to year. The dynamical explanations of these variations are not yet clear. We need to know much more about how the weather systems of the tropics work.

Solar irradiance, and so probably potential evaporation, is less variable from year to year than rainfall. Variations in yield are therefore more likely to be associated with variations in the water balance. Within the rainy months in the humid tropics water stress rarely limits the growth of crops, whose growth rate depends primarily upon intercepted radiation. In the arid tropics the system is less simple. A partial failure of the rains may lead to total failure of the crop, so that yield is not a continuous function of rainfall. Nor do we know enough to understand in detail the ways in which the variations in the water balance affect yield. There are no heuristic

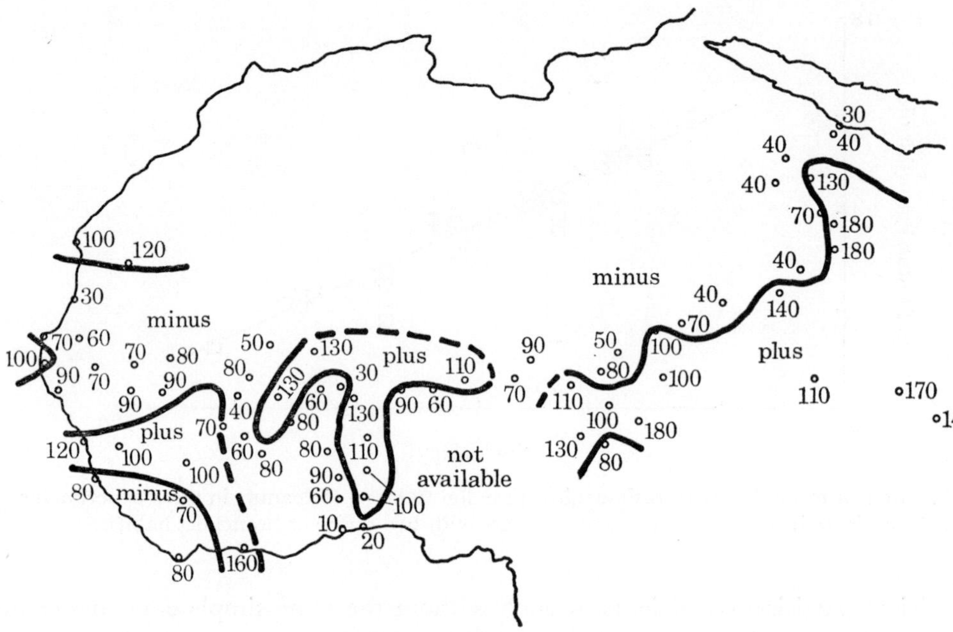

FIGURE 10. A map of West Africa showing the departures of the rainfall for August 1975 from the mean August rainfall, based upon 'Climat' reports.

models and no satisfactory statistical models we can use to assess the consequences for yield of variation in rainfall. The agricultural assessment at the moment depends upon experience, intuition and belief. Within the semi-arid tropics smaller than average total rainfalls appear to decrease arable crop yields. In stable cropping patterns a seasonal total rainfall of about 80 % of the mean sometimes appears to be critical, though grasslands may behave rather differently. The distribution of rainfall is probably as important as the total quantity. Monthly rainfall figures obscure this source of variation in yield. Rainfall totals greater than average seem to be less important.

The monthly rainfalls as reported on 'Climat' can be collected throughout the growing season for any country that reports regularly. The departures of the rainfall for a particular month from the mean for the month, the rainfall anomaly, can be plotted soon after the end of the month (figure 10) and the agricultural significance of the rainfall can be assessed, either qualitatively and intuitively, or statistically.

4. Predictors of seasonal rainfall

Part of the variation of rainfall from year to year is systematic. In the Sahel there has been a statistically unimportant oscillation with a period of more than 5 years (Bunting, Dennett, Elston & Milford 1976). An oscillation implies persistence in some characteristics of the atmosphere from year to year. An oscillation in rainfall, if it continues into the year to come, implies that some part of the rainfall of the season to come is predictable. However, Wright (1971) has listed two rules about such oscillations. The first rule states that some feature of the atmosphere can be found showing a period of the length required to justify a particular hypothesis, and the second states that such an oscillation generally dies shortly after discovery. The first rule is gratifying and both rules seem to be exact.

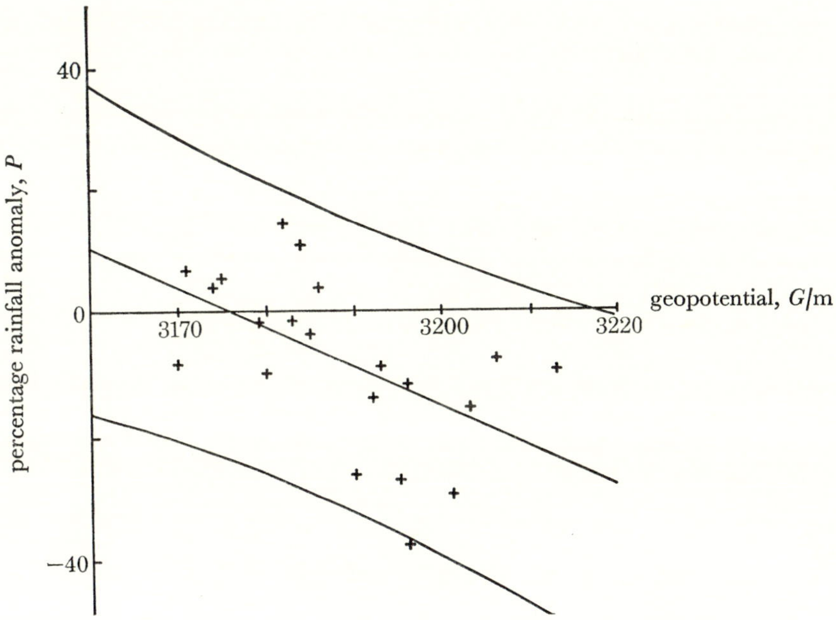

FIGURE 11. The percentage departure from the mean of the rainfall at 50 stations in the Sahel since 1955 plotted against the 700 mbar geopotential at Niamey. A linear regression ($r = -0.55$) and 95% confidence limits have been drawn.

Long range forecasts are now made for a number of countries. The British Meteorological Office releases a 30-day forecast every two weeks. The forecasters have relatively long runs of reliable weather measurements and use sophisticated statistical techniques, including eigenvector methods. Monthly forecast maps for rainfall and temperature are published for the United States of America. The assessment of the reliability of complex forecasts is difficult, but the general tenor of the British forecasts has been reasonable.

Delsi (1973), in a synoptic study of part of Africa, found that the anomaly of the mean monthly 700 mbar (70 kPa) geopotential persisted from June until August. (The geopotential may be taken as the height above sea level at which the barometric pressure reaches a particular value.) The rainfall in West Africa is also persistent from July until September (Bunting, Dennett, Elston & Milford 1975*a*). The 700 mbar geopotential at Niamey can be used to predict the July and August rainfall for the Sahel, provided the atmospheric relationships of the past continue to hold (figure 11). This predictor is statistical. However, the 700 mbar geopotential at Niamey in July depends upon the temperature and humidity of the air of the Inter-Tropical Convergence Zone, and the rainfall of the Sahel depends upon the northerly movement of this moist air. For confidence, the predictor for the July and August rainfall of the Sahel should be based upon the dynamics of the Convergence Zone.

The rainfall of the seasonally arid part of Viti Levu, Fiji, can be predicted from Wright's Southern Oscillation Index (Prasad *et al.* 1975) (figure 12). Rainfall increases with the Southern Oscillation Index. Perhaps the increased pressure gradient between the South Pacific and Australia that occurs when the values of the Index are large increases the advection of moister air and hence rainfall. Certainly the northerly wind component is increased, though there are also differences in the east–west circulation.

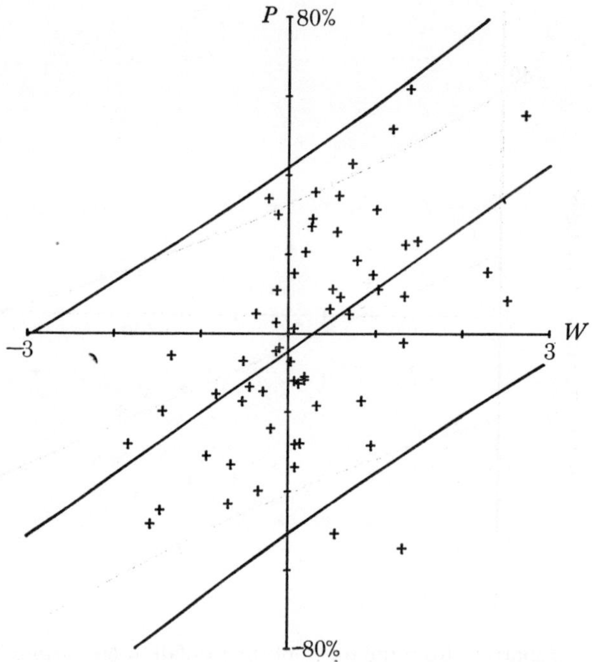

FIGURE 12. The percentage departure from the mean of the August to July rainfall (P) at 5 stations in the sugar cane area of Fiji from 1910 to 1970 plotted against Wright's Southern Oscillation Index (W) for the previous May to July. A linear regression ($r = 0.53$) and 95% confidence limits have been drawn.

The correlation coefficients of the seasonal predictors, though statistically significant, are relatively low. Typically, between 30 and 50% of the variation in rainfall is correlated with the predictor. As a consequence, the confidence limits are wide. We have therefore preferred to use contingency tables specifying the likelihood of wet, average or dry years rather than spuriously precise numerical predictions of the departure of the actual rainfall from the mean.

The chance accumulation of statistical predictors of rainfall would be both tedious and inefficient. Instead the climatological analysis of rainfall must include an analysis of the dynamical meteorology of the seasonally arid tropics. The temporal and spatial variations of rainfall can be analysed by using individual correlation coefficients between stations, correlation matrices and principal component analysis to identify patterns.

The principal component analysis produces a number of components, the eigenvectors, which are linear combinations of stations. The first eigenvector is that combination which accounts for as much variance as possible from the original observations. Successive eigenvectors are extracted from the remaining variance, with the provision that each component is uncorrelated with all previous ones. The analysis has at least two uses. Homogeneous areas can be defined from eigenvectors (see Dyer 1975), and the eigenvectors can be used to describe the large-scale variations of rainfall and to link rainfall with circulation patterns. Pittock (1975) has used these correlation techniques in an admirable study of Australian rainfall. In east and north Australia rainfall and Troup's Southern Oscillation Index are correlated. There are two principal patterns of rainfall anomaly. In the more important pattern, negatively correlated with Troup's Southern Oscillation Index, rainfall is below average in both east and north. In the second pattern, rainfall is above average in the centre of East Australia. These two patterns alone account for 54 % of the rainfall variance. Pittock suggests that much of the observed variation in rainfall may be a more or less random fluctuation, possibly with persistence. It may be possible to predict the likely anomaly from the Southern Oscillation Index.

Table 2. The relation between Wright's Southern Oscillation Index for May to July and annual rainfall at Hyderabad, India, from 1893 to 1972

	rainfall		
	low	average	high
Southern Oscillation Index			
low	18	12	5
high	9	14	22

$\chi^2 = 12.8$; $P < 0.01$.

Extremes of rainfall are associated with Wright's Southern Oscillation Index in India (table 2). When the Index for May to July is small then the annual rainfall is likely to be average or less. The first eigenvector from a principal component analysis of annual rainfall from 1951 to 1974 for 26 stations accounts for 26 % of the total variance of percentage rainfall anomalies. The coefficients are positive at all stations, showing a tendency for all the stations to behave in the same way in any one year. The percentage loadings of the eigenvector show the correlation between station and first eigenvector (figure 13). The relation between rainfall and the first eigenvector is greater in Western Peninsular India, where the loadings reach 82 %. In the east and south the loadings fall to less than 20 %, accounting for less than 4 % of the variance.

The first four eigenvectors can be used to sort the stations into more nearly homogeneous areas (figure 14), which make it easier to establish predictors of rainfall. The first eigenvector is correlated with Wright's Southern Oscillation Index (figure 15) with a correlation coefficient of 0.57 ($P < 0.01$). This eigenvector is very closely related to the average of the Western Peninsular group of 13 stations ($r = 0.98$, $P < 0.001$), and the Southern Oscillation is related to the spatially averaged rainfall of this group of stations.

Clearly there is much work to be done before any predictor of rainfall for Peninsular India, save in the general terms of table 2, can be established. However, also clearly, there is a good chance of establishing such a predictor, using information freely available from 'Climat'.

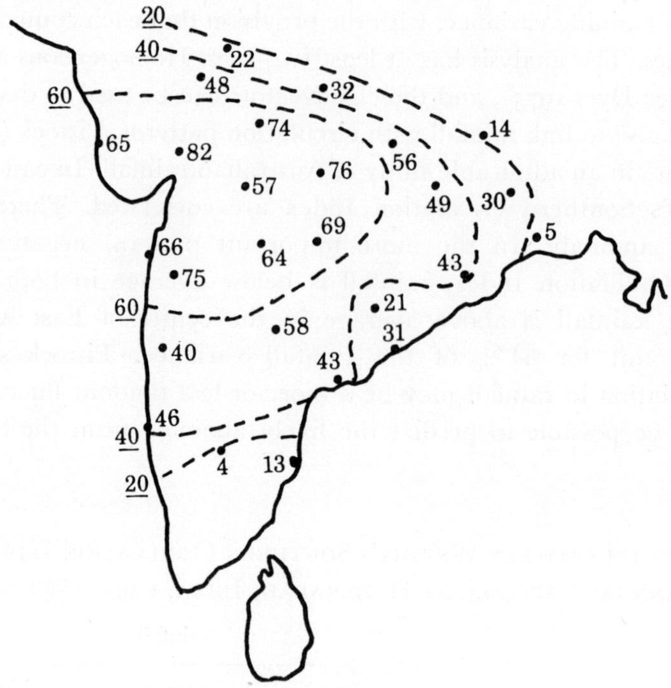

Figure 13. A map of Peninsular India showing the percentage loading of each station on the first eigenvector.

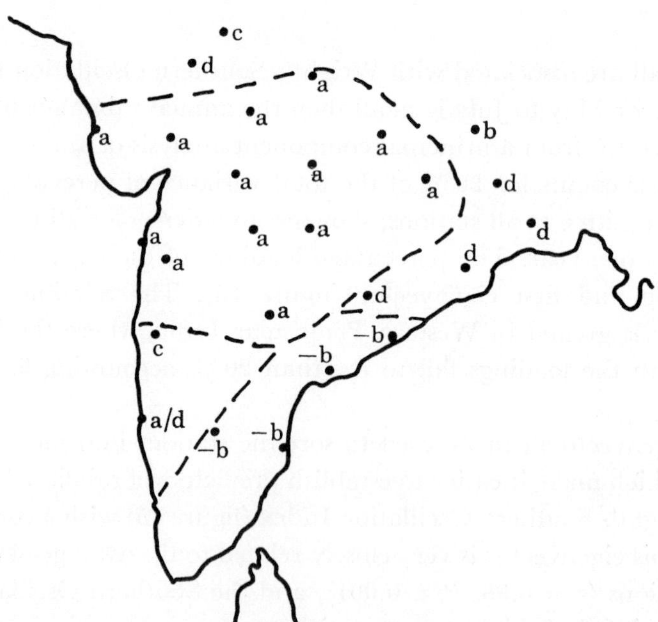

Figure 14. A map of Peninsular India showing the more nearly homogeneous areas of rainfall established from a principal component analysis.

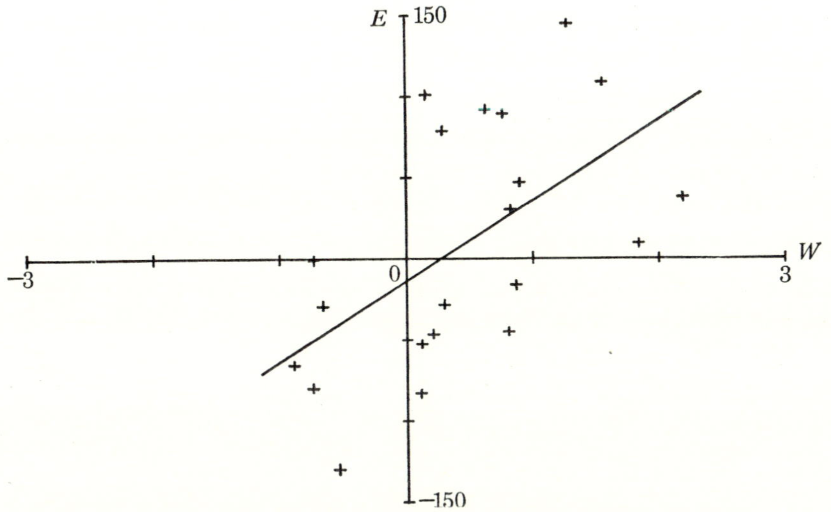

FIGURE 15. The first eigenvector (*E*) of the rainfall from 1951 to 1971 plotted against Wright's Southern Oscillation Index (*W*).

5. The use of forecasts

Subsistence farmers cannot respond directly to forecasts of the character of a season; their farming methods are not sufficiently flexible. The farm calendar is determined by climate and the pattern of crop growth, but the peak demands for human labour are fixed and are very restrictive. Little fertilizer is used; so that any saving from using less in dry seasons is likely to be unimportant in most semi-arid agricultural systems at the moment. It is seldom possible, at present, to substitute one species, or cultivar, for another; but this may change as agriculture becomes more commercial and seed industries develop. Spacing and crop combinations could be altered. Crops could be more widely spaced in dry seasons, though increased vegetative growth may compensate for wider spacing to some extent and so the expected saving of water may be limited. It is possible that some particular combinations of crops, in the multi-cropping systems that are almost universal in the seasonally-arid tropics, are less affected by drought than others. However, crop combinations have been largely neglected in agricultural research programmes and require very much more study.

Governments, on the other hand, can respond to forecasts of the character of a season. If an unduly dry (or wet) season is likely, they can store food and make preparation to distribute it. They can limit cattle numbers, perhaps by controlling access to wells. They can encourage and assist the migration of people away from the most marginal areas.

Perhaps most important of all they may achieve a degree of independence from the weather of one season by developing complementary economic and other exchanges with other climatic areas (Bunting, Dennett, Elston & Milford 1975*b*). Rainfall changes quickly with distance in the tropics. Arid, seasonally arid and humid regions are relatively close to each other. The present national structure in West Africa has developed since the Congress of Berlin. Before that there was a well organized, reciprocal trading system running north–south, linking the drier and wetter regions. The scale of trade certainly depended in part upon the absence of wars and as a consequence varied greatly at different times. Nevertheless the humid south and dry north were linked together; and these arrangements continued and developed through the colonial period. The newer national structure of West Africa has placed administrative and political barriers across these north–south links, isolating the northern countries with their unreliable climates. These countries are politically independent, but they cannot now be economically self-sufficient through agriculture at any but the poorest level, because of their climate, nor are they able to handle the consequences of climatic fluctuation effectively within their own boundaries.

The work described here has been generously supported by the Ministry of Overseas Development. Professor A. H. Bunting and Professor J. L. Monteith, F.R.S., have made major, characteristic contributions to agricultural meteorology and have stimulated, encouraged and corrected us. We are extremely grateful to them. John Goodier of the Agricultural Research Council, Dr P. C. Prasad of the University of the South Pacific, R. A. S. Ratcliffe of the Meteorological Office and P. B. Wright of the University of East Anglia have helped us. We thank them all.

References (Elston & Dennett)

Bunting, A. H., Dennett, M. D., Elston, J. & Milford, J. R. 1975*a* Seasonal rainfall forecasting in West Africa. *Nature, Lond.* **253**, 622–623.
Bunting, A. H., Dennett, M., Elston, J. & Milford, J. 1975*b* Weather and climate in the Sahel. In *Arid lands of sub-Saharan Africa*, appendix: pp. 149–177. Washington: National Academy of Science.
Bunting, A. H., Dennett, M. D., Elston, J. & Milford, J. R. 1976 Rainfall trends in the West African Sahel. *Q. Jl R. met. Soc.* **102**, 59–64.
Delsi, M. 1973 Synoptic circulation in relation to seasonal rainfall and Nile flood in the Sudan. M.Phil. thesis, University of Reading.
Dyer, T. G. J. 1975 The assignment of rainfall stations with homogeneous groups: an application of principal component analysis. *Q. Jl R. met. Soc.* **101**, 1005–1013.
Elston, J., Harkness, C. & McDonald, D. 1976 The effects of *Cercospora* leaf disease on the growth of groundnuts (*Arachis hypogaea*) in Nigeria. *Ann. appl. Biol.* **83**, 39–51.
Elston, J. & Monteith, J. L. in Monteith, J. L. (ed.) 1975 *Vegetation and the atmosphere*, vol. 1. London: Academic Press.
Gates, D. M. & Schmerl, R. B. 1975 *Perspectives of Biophysical Ecology*. Ecological Studies, Vol. 12. New York: Springer-Verlag.
Hadley, G. 1735 Concerning the cause of the general trade-winds. *Phil. Trans. R. Soc. Lond.* **29**, 58–62.
Kibukamusoke, D. E. B. 1958 A note on a more precise method of estimation of the Uganda cotton crop. *Emp. Cott. Gr. Rev.* **35**, 91–99.
Lomas, J. 1972 Forecasting wheat yields from rainfall data in Iran. *W.M.O. Bull.* **21**, 9–14.
Monteith, J. L. 1965 Light distribution and photosynthesis. *Ann. Bot.* **29**, 17–37.
Monteith, J. L. 1972 Solar radiation and productivity in tropical ecosystems. *J. appl. Ecol.* **9**, 747–766.
Pittock, A. B. 1975 Climatic change and the patterns of variation in Australian rainfall. *Search* **6**, 498–504.
Prasad, P. C., Dennett, M. D. & Elston, J. 1975 *A preliminary study of the agricultural climatology of sugar cane yield in Fiji*. Paper read at the Australian Conference on Climate and Climate Change, Monash University.
Wright, P. B. 1971 Quasi-biennial oscillations in the atmosphere. *Weather* **26**, 69–76.

Discussion

J. A. Adedokun (*Atmospheric Physics Group, Department of Physics, Imperial College, London S.W.*7). Elston's and Rainey's papers have mentioned the rôle played by the intertropical convergence zone (I.T.C.Z.) in influencing the Sahel rainfall.

Experience from our work here at Imperial College on West African precipitation shows that a station's position relative to the I.T.C.Z. determines, to a large extent, the amount and duration of its rainfall. The region of maximum convective instability occurs in a narrow band some 7–10° S of the I.T.C.Z. while the atmosphere is rather stable to the north and just south of the I.T.C.Z. Thus maximum convective instability and rain occurs in the Sahel in late July to early August, at which time the southern part (Lagos area) experiences its 'little dry season' (Ireland 1962).

Inter-station (cross) correlation of precipitation in the north–south direction indicates a maximum correlation between Lagos in the south and Sahel stations such as Niamey, Menaka and Kidal to the north, for a two-month lag. Over 20 years of monthly rainfall data (January 1951–December 1970) for Lagos and Niamey, the two-month lag correlation is 0.6. This suggests that the rainfall observed at Lagos can 'foreshadow' that of the Sahel region two months later. This two-month period might be usefully spent in making tactical agricultural plans to match the expected rainfall.

Consideration of a typical wet and a typical dry Sahel year shows that in the wet Sahel year the 'little dry season' extends further north. Moreover, the intensity of the 'little dry season' is related to the strength of the 'Zonal Walker Circulation' (Flohn 1971) and the Southern Oscillation (Walker & Bliss 1932).

May I know if the Southern Oscillation Index used in Elston's paper is the same as that derived by Sir Gilbert Walker in his 'World Weather Studies' or is it a modified form?

References

Flohn, H. 1971 The tropical circulation pattern, *Bonner Meteorologische Abhandlungen*, **15**, 49 pp.
Ireland, A. W. 1962 The little dry season of southern Nigeria. *Nigerian Met. Ser., Tech. Note* No. 24.
Walker, G. T. & Bliss, E. W. 1932 World Weather V. *Mem. R. Met. Soc.* **4**, 53–84.

J. ELSTON. A simple form of the Southern Oscillation Index developed by P. B. Wright has been used. (This has since been published: Wright, P. B. 1975 An index of the Southern Oscillation, *Climatic Research Unit, University of East Anglia, Research Paper* 4.)

S. SANDFORD (*Overseas Development Institute, 10–11 Percy Street, London W1*). Mr Sandford said that in discussing the utility of weather forecasts Dr Elston had concentrated on the use that *cultivators* could make of these. They could also be very useful for *pastoralists* in, for example, the following ways:

(*a*) Enabling them to plan and bring forward their sales of livestock in the face of drought; so that the sales were not delayed, as happens now, until the animals are in very poor condition and prices collapse because of the forced simultaneous sale of livestock in overloaded market and processing facilities.

(*b*) Preventing pastoralists being trapped in isolated grazing areas whence, because of drying up of water points on the route out, neither they nor their livestock could escape when rains are delayed.

(*c*) Enabling, where a system of grazing permits or stock licences are introduced, the quota to be adjusted *in advance* of drought instead of *after* it has started when some damage to the environment has already occurred.

G. STANHILL (*Volcani Centre of Agricultural Research, Israel*). I think it important to point out one serious limitation to the use of weather based statistical and modelling approaches to the study of yield variation. Once the technologically induced yield 'take-off' occurs, weather induced changes in yield become relatively less important even though their absolute size may still increase. Under such conditions the identification and interpretation of the causes of yield variation becomes extremely difficult and complex.

An admittedly non-arid example of this process can be seen in the changing pattern of inter-annual variability of the yield of wheat in England during this century. Since the end of the Second World War, secular causes of yield variation have for the first time become more important than the random variation, indicating that changes in technology are now a more important cause of yield variation than variation in weather.

J. Elston. I wonder if the limitations are as great in developed agricultures as Professor Stanhill believes. Whatever the technology, differences in yield must depend upon the physiology of the crop variety, upon the environment and upon the environment/genotype interactions. There have to be reasons for the differences in yield from year to year, and when these are understood they may be modelled heuristically. Certainly statistical models are intrinsically less good than heuristic ones. However, even now we can account for a good part of the variations in the departures from the regression by variation in weather variables. For example the correlation coefficient between the departures of the yield of wheat in Europe and July and August rainfall is about -0.7, for the period 1952–1971. Some correlation coefficients were even larger for the North American crop.

W. E. Ormerod (*Department of Medical Protozoology, London School of Hygiene and Tropical Medicine, Keppel Street, WC1E 7HT*). Several speakers in this conference have stated that enough scientific and technical information is available for the satisfactory management of the world's semi-arid regions and that the reasons why these regions in Africa, India and Latin America are not so managed are in the main social and economic.

Socio-economic problems are certainly formidable, yet I would like to express my doubt as to whether sufficient scientific information is available or that a satisfactory conceptual framework has been built up to allow us to grasp many of the problems. To illustrate this I would like to put forward two current theories which have not been discussed in this conference but which must nevertheless, if substantiated, be of fundamental importance to management of semi-arid regions. In stating my provenance, as a health worker, I emphasize my inability to assess the validity of these theories; I merely use them to illustrate my point.

The first theory states that the main damage caused by rainfall is, as opposed to that in temporate climates, a result of the high kinetic energy of individual drops (Ellison 1944 *Agric. Eng.* **33**, 491–496; Hudson 1971 *Soil conservation* London, Batsford).

If this theory is valid, at least some of the technology that has been described in the last two days could be dangerous to introduce because of the possible exposure of soils, even for short periods, to high energy rainfall without adequate vegetative cover.

The second theory (Charney *et al.* 1975 *Science, N.Y.* **187**, 434) states that the reflectivity of desert soils exposed by overgrazing can directly cause an inhibition of precipitation, which in the case of the Sahel may be as great as 70%, thereby producing a 'feedback' relationship between the management of vegetation and climate.

It seems that the validity (or otherwise) of these theories is important to our concepts of management. There may be good reasons why these and other theories have not been taken into consideration, but their neglect and that of other seemingly important concepts leads me to question whether we have sufficient scientific information to enable us to transfer, with confidence, an empirical technology from one region to another in the face of great socio-economic difficulty.

J. Elston. Tropical agronomists are well aware of the damage that rainfall can cause to bare soils and that cropping systems are being designed to minimize it. Lawes has been testing such systems at Samaru in the early 1960s and they have since been developed further. Professor Charney's theory is nice and probably at least partly true. It is not possible to test it with a real experiment, but it can be assessed gradually as more measurements are made at the desert margins.

M. Mansell-Moullin (*Old Hatch, Lower Farm Road, Effingham, Surrey*). Previous speakers have mentioned the need for both more basic water resources data for development schemes in the semi-arid regions and also the greater transfer of knowledge between English and French-speaking scientists in this field. I strongly support both views.

In these regions the hydrological characteristics of the water resources fall into two groups – those which can be defined reasonably well from observations covering 2–3 years and those requiring much longer, say 10–50 years. Sufficient data on the former can be obtained during project investigations but the latter must depend mainly on existing records, usually transposed from elsewhere. Group 1 contains the more stable elements, e.g. the general climate, ground-water conditions, chemical water quality, etc. and group 2 mainly those related closely with precipitation, e.g. storm intensities, monthly and annual precipitation variations, runoff characteristics, sediment runoff, etc. Frequently the only information available on the items in group 2 is the knowledge of the local inhabitants, if any.

The limited, world-wide information on group 2 is seldom in a transposable form. I wish to recommend that an appropriate international organization should tackle this problem by (*a*) identifying those regions and countries where important water development schemes are having to be based on scarce data, (*b*) determining where there are reliable, long-term data from comparable areas and (*c*) reviewing and generalizing that data so that it can be transposed readily. The programme should be regarded as a pooling of present knowledge, rather than as research.

Phil. Trans. R. Soc. Lond. A. **278**, 611–614 (1977) [611]
Printed in Great Britain

Review and prospect: where do we go from here?

By A. H. Bunting
University of Reading

I cannot presume, in the time at my disposal, to sum up so diverse and excellent a series of addresses as we have heard during these two days. I propose instead to review what seem to me to be the most significant issues they raise affecting action for the future of the arid regions, with particular reference to those which form parts of the less-developed countries of the world.

Development and governments

To develop resources is an exclusive prerogative of sovereign, independent governments. Other agencies and individuals have authority to develop resources only if governments permit them so to do. Though they may not always be fully effective in practice, the declared policies of most, if not all, governments, particularly in the less-developed countries which contain many of the arid lands we have been discussing, commit them to direct the development of resources to the interest of the nation as a whole. Now in the last analysis, development is about people; and most of the people of developing countries are members of societies or communities in which agriculture and stock-raising are the chief, and often the only, processes which use resources productively. My first conclusion, therefore, is that resource development in arid lands is an integral part of the policy and action of governments for agricultural and rural development, and that it is very far from an exclusively academic, ecological or environmental preserve.

Very few of us here have been directly involved as citizens in the development of arid regions. The nearest most of us have got to the action is in the rôles of servants and advisers of governments, or of the national and international agencies that cooperate with them. Others of us are research workers, teachers and communicators who increase and transmit knowledge and understanding. But few of us have borne the heat and burdens of drought and disaster on our own backs. For me, that is a particularly humbling thought.

Population

Some of us are deeply and sincerely concerned with the growth of population in arid and other developing regions of other people's countries. To a large extent it is inevitable that populations should grow at an increasing rate in the developing world at the present time. As preventive and curative medicine become more effective and as food supplies per head increase, or at least do not decrease (which in spite of well-publicized droughts and disasters, is what has happened, by and large, for 30 years past) more babies survive, average expectations of life become greater and the growth rates of populations increase – even if birth rates fall, as they are tending to do in many of the poorer countries. These life-increasing processes are regarded by most human beings as morally good, and they are not affected by family limitation programmes – of which let me hasten to say, I am a long-term financially committed supporter. The main lesson I draw from these considerations is that while many of us will wish to do all we can to help those governments that wish to do so to decrease birth rates, by whatever means their people can accept and afford, we must also help them to prepare for continuing population growth, for perhaps a century to come. A population policy, still less a family limitation policy cannot serve as a cheap substitute for a viable national (and especially agricultural and rural) development policy.

A subsidiary but nevertheless salutary conclusion is that we should think more carefully about how far a world food or protein shortage, which would be expected to kill many babies and young children, is intellectually compatible with a world population 'explosion' which develops because more of these same babies and young children are surviving. There must be a limit to the extent to which we can have it both ways.

Socio-economic factors in life systems

We have, as is usual in such discussions as these, met our old friends the 'socio-economic factors'. A distinguished but orthodox Fellow of the Society once described the early Unitarian Church as a feather bed for a falling Christian: 'socio-economic factors' have all too often provided a feather bed for a falling agronomist. When his prized new variety, or his 'package of improved practices', the products of devoted, even impeccable, investigation, fail to win instant acceptance, his first response is usually to blame the extension service; and when that in its turn has to be exonerated, he turns to socio-economic factors, which he may conceive as having to do with marriage customs, fertility rites, totems, taboos, and such-like apparently, non-agricultural matters. In this way he dexterously puts the problem not only off his personal plate but also beyond the reach of objective study, for lack of suitably equipped investigators.

Fortunately, during the last 35 years, such investigators have begun to join the action. From them we have learnt that many of the socio-economic relations of traditional agriculture are important for us because they affect decisions about the volume and distribution of scarce resources of land, labour, time and capital between different and competing sectors of the life-systems of rural societies.

The key word is 'life-systems'. The traditional farmer or herder of stock is not like a Berkshire farmer, who can devote virtually all his resources to his farm business: he and his family have to do many other things in order to live and survive, and they have to share out their resources accordingly.

Instability and change in arid regions

The next point which seems to me to be important for the future is the inherent instability of systems of resource-use in arid regions. They are unstable because the supply of the limiting resource, water, is distributed unevenly and unpredictably in both space and time. This sets limits to the extent to which life and human activity can be stabilized, unless resources can be stored or supplemented, or unless people, or resources, or both, can move or be moved, at the right time and speed, from place to place.

Indeed, in arid regions, change is characteristic of both the situation and of the remedies for it: little or nothing can be stable or sustained. How, in such situations, can one define and measure the 'carrying capacity' of the environment? In these circumstances, there are and have been continuous and substantial changes in the type and productivity of vegetation; and in many cases these changes, reinforced by inappropriate methods of resource-use, have made ecosystems less productive.

To add from outside to the resources of the life-system in an unstable environment often seems to make matters worse. A society of nomads, who have, apparently for generations, managed grass and water along their routes so that they and their resources survive, will apparently grossly mismanage and degrade the pasture within kilometres of a new borehole provided by government. Why? Is this a special case of the general one, that in these circumstances the animals are privately owned while the resource is publicly owned? If so, why did this problem not present itself in the bad old days?

Productive and unproductive change

Not all change, however, makes the environment less productive: there are old fixed dunes, now carrying settled agriculture, hundreds of kilometres south of the true desert in Kordofan; annual crops on land cleared from *Acacia* woodland with tall grass on the Sudan clay plains can be managed so as to use water more productively than the wild vegetation; net production of cropland and pasture in England is generally far greater than that of the deciduous broad leafed woodland it has replaced. Change processes are of many sorts, and not all are necessarily deleterious. We may indeed expect that during the more humid period which is now returning to the Sahel, and which will probably continue, with ups and downs, for ten or more years, the pressure on land in the broad ecotone between the desert and the sown will decrease, and that plant cover and productivity will improve, except where truly irreversible damage has been done.

New technology

To be effective in lessening general poverty, agricultural and rural development, in less – as well as more-arid areas, must embrace very large numbers of people. The average improvement in the productive use of resources need not be spectacularly large; but even modest improvements are not likely to be realized unless they are part of a general and complex set of changes extending far beyond the metaphorical farm gate. Physical infrastructure, technical installations, commercial input and output channels (whether publicly or privately operated is less important than that they work) including marketing, storage, transport and price management, are all necessary parts of the deal. Through them, the agricultural and the non-agricultural parts of the society, and of the economy, are linked and indeed to an extent polarized: exchange between them benefits both. All this rests on investment decisions by governments and their advisers, who may not be particularly willing to make major investments in the least promising or the least reliable areas.

Technical means of using resources in arid lands more productively than now can surely be found, but without the rest of the complex set of changes to which I have just referred, not many farmers will be able to adopt them. Moreover, to be accepted, the new means must be within the technical grasp of rural people, they must lessen their felt difficulties, and they must do this at a price they can afford without undue risk. Often, the limiting factor is not technical at all – in the sixties a new all-weather road between Kano and Zaria in Nigeria led to great increases in the production of sugar cane and vegetables for sale, without benefit of research, education or extension.

Ecological complementarity and forecasting

In many instances, particularly in West Africa, arid regions are not far removed from more favoured ones. The traditional systems of resource use are based on movement, and on exploiting ecological complementarity between stock- and grain-producing regions. Indeed it must seldom be possible for a Sahelian state to act within its own boundaries against fluctuations in climate: to do this successfully it will need appropriate complementary relations with wetter regions. It is when these relations are inadequate or impossible (for political or administrative reasons) that even a moderate drought can lead to a large-scale disaster.

Forecasting, whether of an impending spell of dry years or of an impending dry season, appears to be sufficiently reliable to offer advance warning to governments so that they can look to their ecological complementarities – and to the storage, transport and distribution systems necessary to realize them in time.

Prospects of a better age

Perhaps we should next remember that in the arid lands people are changing also. The transistor radio brings tidings of the wider world to the most remote encampments and hamlets, and tells their inhabitants they that are not obliged for ever to continue in the old ways. The Fulani are settling in the high pastures of West-Central Cameroon; the young and the educated are leaving the harsh environments, particularly where oil and other minerals, in which so many arid lands are rich, are changing the patterns of economic activity and opening up new prospects. We may expect this sort of movement to continue and to accelerate – as it has done from marginal lands in many countries in the past. As development proceeds, the ordinary people of the arid lands, for good or ill, will take their own decisions about their future.

In the two days of this meeting many people with diverse interests and experiences have shared their knowledge, their ideas and their hopes. We have met against the background of the widespread droughts of 1972–3, but at a time when the immediate urgencies are less, to consider the possibilities for the future, when drought returns again, as it surely will, to the Sahel and other arid lands. I believe we now share a common conviction that much can be done, and a determination to help to do it. For this not only we ourselves, but the millions in the arid lands on whom the main burden will lie, will join in thanking this Society, our generous host, for making this notable occasion possible.